T0237463

SpringerBriefs in Finance

More information about this series at http://www.springer.com/series/10282

Othmane Kettani · Adil Reghai

Financial Models
in Production

 Springer

Othmane Kettani
Quantitative Research Equity
Natixis
Paris, France

Adil Reghai
Quantitative Research Equity
Natixis
Paris, France

ISSN 2193-1720   ISSN 2193-1739   (electronic)
SpringerBriefs in Finance
ISBN 978-3-030-57495-6   ISBN 978-3-030-57496-3   (eBook)
https://doi.org/10.1007/978-3-030-57496-3

Mathematics Subject Classification: C51, 91B70, 91B28

© The Author(s), under exclusive license to Springer Nature Switzerland AG 2020
This work is subject to copyright. All rights are solely and exclusively licensed by the Publisher, whether the whole or part of the material is concerned, specifically the rights of translation, reprinting, reuse of illustrations, recitation, broadcasting, reproduction on microfilms or in any other physical way, and transmission or information storage and retrieval, electronic adaptation, computer software, or by similar or dissimilar methodology now known or hereafter developed.
The use of general descriptive names, registered names, trademarks, service marks, etc. in this publication does not imply, even in the absence of a specific statement, that such names are exempt from the relevant protective laws and regulations and therefore free for general use.
The publisher, the authors and the editors are safe to assume that the advice and information in this book are believed to be true and accurate at the date of publication. Neither the publisher nor the authors or the editors give a warranty, expressed or implied, with respect to the material contained herein or for any errors or omissions that may have been made. The publisher remains neutral with regard to jurisdictional claims in published maps and institutional affiliations.

This Springer imprint is published by the registered company Springer Nature Switzerland AG
The registered company address is: Gewerbestrasse 11, 6330 Cham, Switzerland

Foreword

Options theory has come to us in so many ways. Beyond numerous technicalities, one is faced with unknown, possibly time varying, data generating processes and hedging issues in possibly incomplete markets. Extensive work has been done in these directions, but it is seldomly applicable in a real-world context. Therefore, over the years, an engineering approach has emerged, aiming at reconciling practicalities and rigorous approaches. Rigour derives from brevity, clarity and precise descriptions of pricing and risk management procedures. Leveraging on Adil Reghai's "Quantitative Finance: Back to Basic Principles", this book provides a self-contained investigation of volatility dynamics and the intersection of option pricing and risk management, including an explanation of P&L. This work has unique and quite distinctive features: it takes the reader straight to the points addressed. Being built on real-life experience, it is full of innovative views and practical results not easily accessible elsewhere. Last but not least, it is a testimony to the vibrant intellectual climate within the quantitative finance community. For the above reasons, Adil Reghai and Othmane Kettani's book will be of interest to a wide audience, ranging from students in mathematical finance programmes to seasoned academics and quantitative engineers. Besides being useful and helpful, it sparks curiosity and brings the pleasure of learning as one moves through its different chapters. I had the opportunity to meet with Adil and Othmane in recent years. It is my honour and privilege to once more enjoy their expertise.

Jean-Paul Laurent
Université Paris 1
Panthéon-Sorbonne
Paris, France

Preface

In this book, we show how to put in production models with important features at the minimum cost of implementation and the maximum robustness and control.

The idea is that one can implement production level models based on elementary and robust building blocks.

The main building blocks for equity derivatives, which are our subject here, are the following:

- Black–Scholes formula,
- Implied volatility calculator,
- Local volatility pricer,
- A new breed of stochastic volatility.

With these blocks and a lot of astute, one can build all the necessary models in order to risk manage all the costs involved in the options fabrication within the world of equity derivatives and hybrids.

Paris, France

Othmane Kettani
Adil Reghai

Contents

1 General Introduction 1
   1.1 Role of the Quants 1
   1.2 PnL Explanation 4
   1.3 Exploring Models 4
   References 5

2 Black & Scholes (BS) Model 7
   2.1 Closed Form Formula Revisited 8
   2.2 Smile Representation 10
   2.3 Summary of the Chapter: Key Messages 13
   References 13

3 Local Volatility Model 15
   3.1 Some Stylized Facts About of the Local Volatility 17
   3.2 Other Properties of the Local Volatility 19
      3.2.1 The Skew Stickiness Ratio (SSR) 19
      3.2.2 The Local Volatility Forward Skew 21
      3.2.3 The Local Volatility SSR in the Future 24
   3.3 Local Volatility Model: A Pricing Example 25
   3.4 Local Volatility PnL Explanation 26
      3.4.1 Derivation of the P&L Equation 26
      3.4.2 Another Simpler Derivation 28
      3.4.3 PnL Explanation Factors 29
      3.4.4 How Do We Test the Hedging Hypothesis
          in Practice? 29
   3.5 The Standard Theta 32
      3.5.1 Decomposition of the Standard Theta:
          The Exotic Theta 32
      3.5.2 Decomposition of the Standard Theta:
          The Vanilla Theta 33

3.6   Typical Portfolio   34
3.7   Generalize This Approach to Multi Assets and Quanto Effect   34
3.8   Summary of the Chapter: Key Messages   35
References   37

4   Market Model P&L Explain   39
4.1   Theory   40
      4.1.1   Derivation of the Adjustment   40
      4.1.2   Computation of the Second Order Derivative   42
4.2   Focusing on the Vanna   43
      4.2.1   Covariances Spread Computation   43
      4.2.2   Vanna Computation   44
4.3   Analysis of a Real Case Example: Case of the Autocall   45
      4.3.1   The Payoff   45
      4.3.2   Vanna Adjustment for the European PDI Up&Out   47
      4.3.3   Vanna Adjustment for the Coupons   49
4.4   Summary of the Chapter: Key Messages   51
References   53

Annex: Derivation of the q_{KT} Expression in the Case of a PDI UO   55

Bibliography   59

Index   61

List of Figures

Fig. 3.1   Representation of a typical backbone ATM-volatility   17
Fig. 3.2   Representation of the at the money volatility slope.   17
Fig. 3.3   Representation of the at the money curvature   18
Fig. 3.4   Representation of a full implied volatility surface   18
Fig. 3.5   Representation of the local volatility surface.   19
Fig. 3.6   Term-structure of SSR. $\alpha = 0.5$ 20
Fig. 3.7   Term-structure of SSR. $\alpha = 0.3$ 21
Fig. 3.8   Term-structure of SSR. $\alpha = 0.7$ 21
Fig. 3.9   Forward skew in the LV Model. Parameters of the plot:
            S0 = 100%, K = 90.48%, r = 1%, q = 3%, T = 5,
            Bump = 1%   23
Fig. 3.10   Forward covariance between log-spots and implied volatility in
            the LV model. Parameters of the plot: S0 = 100%,
            K = 90.48%, r = 1%, q = 3%, T = 5, Bump = 1%.   24
Fig. 3.11   Forward SSR in LV Model. Parameters of the plot:
            S0 = 100%, K = 90.48%, r = 1%, q = 3%, T = 10,
            Bump = 1%   25
Fig. 4.1   Autocall payoff Cash flows as function of the
            underlying level   46
Fig. 4.2 $q(DB_N, N)$ for the PDI. DI barrier at 60%. Maturity = 5y.
            The point of implied volatility of interest is (60%, 5y).   48
Fig. 4.3   Vega KT of the PDI and the PDI UO with respect to the point
            of implied volatility (60%, 5y)   48
Fig. 4.4   Vanna adjustment for the short 5-year PDI UO position
            (strike = 100%, barrier = 60%) with annual autocall at 100%.
            Vanna adjustment is displayed in bps   49
Fig. 4.5 $q(AB_i, i)$ for the coupon UO Ti = 2y. Autocall barrier at 100%.
            The point of implied volatility bumped is (100%, 2y)   50

Fig. 4.6   Vega KT of the European digit and the Coupon UO with
            respect to the point of implied volatility (100%, 2y).   50
Fig. 4.7   Vanna adjustment for the coupons UO (spot = 100%, autocall
            barriers = 100%) with annual autocall. Vanna adjustment is
            displayed in bps   51

Chapter 1
General Introduction

Abstract This general introduction highlights the role of the quants and the main challenges they have to overcome when pricing financial derivatives. We state the main equation driving the economics of the quant role and give a detailed explanation of each term it contains. We also clarify the material that the reader may find in this book as well as the prerequisites needed to understand its content.

Keywords EVA · Cost of capital · P&L

1.1 Role of the Quants

The role of the quants is difficult to define. It is constantly changing and the job is adapting to the situation of each bank but also to the global economic and regulatory environment.

What is common with all teams is that quants spend their time between an exercise of support and production in one hand and another of exploration in the other hand.

Support and production or exploitation is the daily help of the business in order to provide tools and analysis to the different operators in the markets. This is typically oriented towards trading teams. They need to make sure that they are using the right model. They are fine tuning the parameters in such a way that within a given market and a given structure they are doing the right thing. A quant in these operations needs be flexible and agile. The main quality is intelligence, good communication and team work. Sometimes they are called desk quants or commandos.

Exploration or economic intelligence follows a different rhythm. Projects are prioritised for longer times. One needs to implement solid algorithms that adapts to the usage. The first exercise of support and production helps reinforce this exercise by providing operational experience on all situations. Exploration is also a time for thinking and getting more insight about the practice. It is also a research approach whereby one must be ready to fail. Even worse, research without failing is not research.

© The Author(s), under exclusive license to Springer Nature Switzerland AG 2020
O. Kettani and A. Reghai, *Financial Models in Production*,
SpringerBriefs in Finance,
https://doi.org/10.1007/978-3-030-57496-3_1

These two exercises need to be well balanced. An excess of one of them makes the team weaker. And a good balance between aspirations and skills is a dynamic exercise of research management.

This being said, there is one equation that drives the economics of the quant work.

$$EVA = PnL - Costs - Cost\ of\ Capital \qquad (1.1)$$

EVA is the equity added value to the business. It is a key indicator that is followed by the top management and the financial analysts all along. Positive EVA is a sign of good management and a good business.

Costs mean the cost of doing business. This includes direct costs such as salaries of business operators and their support. Quants as well as ITs are part of it. It also includes indirect costs which are composed of operations, back office, middle office, risks, compliance, legal etc. All these teams participate to the success of a given business.

Cost of Capital is linked to the regulation in place which stipulates that some capital must be put in front of the business to face all sorts of risks. The main risks that are facing a bank are credit risk, market risk and liquidity risk.

Quants adapt to this equation offering enough developments and talents to cover each topic.

PnL The main objective is to be able to propose the right modelling framework for our traders to risk manage properly their clients' products. Unlike physics, these models tend to evolve depending on the global context (which products, which margins, which liquidity). Quants work a lot to propose a wider product catalogue and diversify the activities. Their contribution is twofold. First, increase the innovation, and second, preserve the PnL offering good tools for pricing and risk management. This book will present the most important tools, production-proof, developed for pricing and risk managing, which will definitely help preserve the business.

Costs Here quants are using the latest statistical and now called machine learning techniques in order to improve the operations. The best successes concern the automatic treatment of market data to feed the systems (volatility most importantly but also interest rates, dividends, correlation etc.). Another spectacular achievement is the usage of machine learning in order to improve the automatic risk management of a derivative book. If we look at the previous examples, the first one was helping feed the system and the second is helping hedging operations. To continue, one important usage of machine learning is the anomaly detection which helps dramatically the overall quality of the whole process. These three applications of machine learning improve the quality and reduce the cost of the operations.

Cost of capital Before all, one needs to compute the cost of capital which involves a tremendous amount of calculations. This is true for the standard approach in the FRTB regulation, especially for the default risk charge where one needs to compute the behaviour of the portfolio for every single stock going to default. This is even worse for the advanced methodology where one needs to compute millions of scenarios. Quants have a great role rethinking the system's architecture and designing new types of algorithms to perform such challenges. Besides, once the calculation is done, quants can help understand the profitability of the different businesses as well as help optimising them.

Quants went from a world of pricing and designing models for this purpose to proposing hedges, computing regulatory indicators in order to have the exact vision of business profitability.

In this book, we explore a new frontier where we deeply study the PnL evolution of a model in order to understand what comes from the market, what comes from the position and what comes from the model. Thus, we will propose a frugal and pragmatic approach to improve the whole management of a derivatives book. It is the result of a collective work with trading, Risk, IT and quants. The proposed innovations are opening new avenues to make a business profitable, transparent and secure.

This book is a hands-on book that provides the reader with the means to understand how financial models are actually implemented and used in production, on a daily basis, for pricing and risk-management purposes. The authors show how to put in production models with important features at the minimum cost of implementation and the maximum robustness and control. Thanks to a robust implementation, we show how to explain the P&L for real models in real situations. This is achieved by extending classical models and improving them in order to account for complex features.

Although the book is primarily meant to market practitioners, it might be of interest for students in Quantitative Finance/Mathematical Finance Masters as well. In the end of each chapter, the reader may find a summary with the key messages (chapter's take away). In addition to that, the authors designed some exercises to consolidate the reader's understanding. Of course, to be able to fully grasp the content of the book, the reader needs to have some general background in the following areas: stochastic calculus, pricing of financial derivatives, quantitative finance. However, it is less mathematically demanding than some other books on the subject (cf. Guyon and Henry-Labordère 2013) as the authors' focus is rather production oriented where, based on some well-chosen results, they provide practical direction for risk managing exotic financial derivatives using classical models.

An advanced reader may find a number of similarities with the following references Bergomi (2015, 2018). The similarities concern essentially the Chap. 3 of the book (specifically parts 3.2, 3.4 and 3.5) where we present some of the results the reader may find in the above-mentioned references. Indeed, the authors recall some of these results in this chapter as a pre-requisite to understand the risk managing approach in Chap. 4. We do not provide any proofs of these results and advise the interested reader to refer to these references for further mathematical details. Our approach is simply to introduce some required notions and concepts [such as the local volatility properties (SSR)] in order to prepare the reader for the methodology presented in the final chapter.

1.2   PnL Explanation

One essential objective in modelling is to make it to production. The information about the model, brought by using it on a daily basis in different market regimes and contexts, is tantamount and shows how it fits the market.

G. Box, a well-known statistician, said that "All models are wrong but some are useful". Our motto is a bit different. It says that "All models are useful but they all come with limitations".

Based on this remark one should always have in mind the scope where it is possible to use a model and especially have a clear vision when the model fails and be prepared for an alternative.

Models bring numbers and hide the true reality of the risk. It is essential to understand how each model functions, how it is used and the influence of the market where it is engaged.

We promote an incremental approach going from simple to more complex models. We try to use a model in practice and learn what is missing in order to improve it and add the necessary component to make this model work under the conditions of use.

To perform this simple program one needs to have a deep grasp of how models work in practice, under what market conditions, but also and most importantly, how they are used by practitioners.

1.3   Exploring Models

In a previous book, (Reghai 2015), we showed when it is possible to use a Black Scholes model (constant gamma sign exposure). We also saw that local volatility could take into account the changing sign of the gamma. We also introduced a toxicity index which indicated if one could do with a local volatility model or should use a more sophisticated approach. We observed that for two given models fitting the smile, a parametric local volatility and a stochastic volatility, the difference in prices came from differences on the dynamic of the smile. The approach was pedagogical and showing parametric cases where it is possible to build a parametric local volatility model that is consistent with a stochastic volatility.

In a nutshell, we showed that the two models M1 and M2 gave similar smiles and gave different dynamics **with different costs of smile. By 'costs of smile', we mean the actual costs, due to the smile, that the trader incurs while hedging the financial derivative. In concrete terms, these costs appear in the P&L equation through sensitivities such as Vega, Vanna and Volga (to a certain extent).**

M1 stationary parametric local volatility,

$$\frac{dS_t}{S_t} = \sigma(S_t)dW_t$$

$$\sigma(S_t)^2 = \alpha^2 \ln^2\left(\frac{S_t}{S_0}\right) + 2\alpha\rho\sigma_0 \ln\left(\frac{S_t}{S_0}\right) + \sigma_0^2$$

M2 stochastic volatility model as a special case of SABR model (cf. Hagan 2002),

$$\frac{dS_t}{S_t} = \sigma_t dW_t$$

$$\frac{d\sigma_t}{\sigma_t} = \alpha\left(\rho dW_t + \sqrt{1 - \rho^2}dW_t^\perp\right)$$

where the Brownians $\left(dW_t, dW_t^\perp\right)$ are independent.

In the following sections, we tackle production problems with realistic smiles and common production questions.

References

Bergomi L (2015) Stochastic volatility modeling. Wiley

Bergomi L (2018) Selected papers. Retrieved from lorenzobergomi.com

Guyon J, Henry-Labordère P (2013) Nonlinear option pricing. Financial mathematics series. Chapman & Hall/CRC Press

Hagan KL (2002) Managing smile risk. Wilmott Mag 84–108

Reghai A (2015) Quantitative finance: back to basics. Palgrave MacMillan

Chapter 2
Black & Scholes (BS) Model

Abstract The Black & Scholes model is still topical in the market finance industry. The price of a European option is yet computed with a simple closed-form expression that depends on market observables. In this chapter, we do not present this model but rather focus on how it is used nowadays in production. That is to say, how practitioners use it to imply volatilities from market quotes. We revisit the closed-form formula and present a different methodology to compute it in order to overcome the numerical challenges inherent to volatility implicitation. Building on that, we explain why the Black & Scholes model fails in pricing and risk managing financial derivatives, while still being consistent with the market, and present an alternative model, at the end of the chapter, to construct a flexible and consistent market smile.

Keywords Black-Scholes · Volatility implicitation · Volatility smile · Volatility term-structure

This chapter is mainly devoted to the BS model and its use in production. This model is well known by every practitioner. It has this great advantage of being analytical i.e. it comes with a closed form formula. It is of great help when using it in a daily trading activity. However, there is confusion between the closed form formula and the model assumption. Indeed, on one hand the model assumption stipulates that the return of the underlying are log normal (one must understand that there is no smile generated by this model). On the other hand, if you apply these assumptions to a classical vanilla option i.e. European call or put option, you obtain a closed form formula.

Most banks do not hedge and risk manage with BS, but rather use it to provide the inputs to calibrate more sophisticated models such Local Volatility (or Local Stochastic Volatility), which are the models that are actually used for pricing and risk management purposes. The main use of the BS model, production-wise, is volatility implicitation from available market prices. That's why we present in the following a smart way to use it in order to overcome numerical challenges.

© The Author(s), under exclusive license to Springer Nature Switzerland AG 2020
O. Kettani and A. Reghai, *Financial Models in Production*,
SpringerBriefs in Finance,
https://doi.org/10.1007/978-3-030-57496-3_2

2.1 Closed Form Formula Revisited

Having a closed form formula is the exception in the world of derivatives modelling and permits fast, accurate and robust pricing and risk management. These characteristics are essential and highly desirable for every model put in production.

This certainly contributed to the success of using the Black & Scholes model by traders. However, we believe that the real reason of its wide success is its handiness and simplicity, particularly the fact that all parameters are given by the theory of hedging and there is only one free parameter to control to obtain the option's price. This essential parameter is the **volatility**.

Here we are, with a model applied to vanilla options which comes with analytical formula and where only one parameter i.e. the volatility must be set in order to use the model.

To back out the level of implied volatility from prices we use numerical algorithms. There is no exact formula for this exercise. A closed formula does not exist but there are algorithms that do the job.

There are many algorithms that perform the task. In practice (Jäckel 2006) provides machine precision accuracy and is very robust while still being fast. The difficulty of implying the volatility is the fact we are dealing with a super smooth function of the volatility and when the price is very small we have first order derivatives which are nil. It is therefore difficult to use Newton type formula where we divide by the first order derivative.

We Present a new strategy of implying the volatility and perform comparison analysis with Jäckel (2006) approach (Table 2.1).

We have introduced a new implementation that performs very well. It is both accurate and robust. On one hand, it relies on a good initial guess (Stefanica 2017) and on the other hand, it is based on a classical search algorithm zbrac and zbrent as in Chap. 9 in Press and Teukolsky (2007). What is **different** is the implementation of the Black & Scholes formula. The classical implementation, which is a difference of two real numbers, shows rounding error problems during the search, especially when the option price is small. We use instead an implementation that is the sum of positive contributions. It has the nice property of remaining positive whichever the discretisation.

Table 2.1 Performance of the algorithms (Jäckel and Reghaï&Kettani) algorithm using F = 100, K = 150, T = 1.0 with a relative error of 10e−8

Original volatility (%)	Method	Implied volatility	Time (ns)
64	Jäckel	0.640000000000002	1005
	Reghaï&Kettani	0.640000000000160	117
16	Jäckel	0.160000000000002	1323
	Reghaï&Kettani	0.160000000000180	117
4	Jäckel	0.039999999999999	1519
	Reghaï&Kettani	0.039999999997073	348

Mathematically speaking, we went from the classical Black Scholes formula (undiscounted price)

$$Black_{Call} = FN(d_1) - KN(d_2) \tag{2.1}$$

where:

$$d_1 = \frac{\ln\left(\frac{F}{K}\right) + \frac{1}{2}\sigma^2 T}{\sigma\sqrt{T}}, d_2 = \frac{\ln\left(\frac{F}{K}\right) - \frac{1}{2}\sigma^2 T}{\sigma\sqrt{T}}$$

- T is the option maturity,
- F is the forward of the underlying,
- K is the strike of the option,
- And σ is the option's volatility.

to a **different** implementation, which replaces the two cancelling terms with only positive contribution. Indeed, our implementation of the Black & Scholes price is the sum of the intrinsic value and the time value: two positive terms.

$$Black_{Call} = [F - K]^+ + F\sqrt{T} \int_0^\sigma n(d_1(u)) du \tag{2.2}$$

where $d_1(u) = \frac{\ln\left(\frac{F}{K}\right) + \frac{1}{2}u^2 T}{u\sqrt{T}}$ and n being the standard Gaussian density function.

This formula is nothing else than saying that $f(\sigma) = f(0) + \int_0^\sigma f'(u) du$ for any regular function f. The Black & Scholes price is then equal to the intrinsic value plus the Vega $F\sqrt{T}n(d_1(u))$ contribution from 0 to the given level of volatility σ. The key element for robustness is that this price implementation is always the sum of positive contributions. We eliminated the rounding error problem using this little trick.

We can even improve the implementation using the following expression:

$$Black_{Call} = (F - K)N(d_2) + F \int_{d_2}^{d_1} n(u) du \tag{2.3}$$

In the case of the classical implementation of the Black & Scholes formula, one uses approximations for the cumulative normal. In this new formulation we shall use a quadrature algorithm **Gauss-Legendre** described in Chap. 4 in Press and Teukolsky (2007).

It stipulates that for any function f one can approximate the integral as follows:

$$\int_{-1}^{+1} f(x)dx \approx \sum_{i=1}^{n} w_i f(x_i)$$

The approximation is exact for all polynomials of degree $2n - 1$ or less.

This is to say that the price practical implementation is a weighted sum of Vega contributions at a well-chosen set of points. In practice 10 points seems to be the right fine tuning parameter for achieving machine precision calculation.

This small change in implementation generates an extremely fast, accurate and robust implied volatility calculator. Nothing else is changed and yet we obtain extremely good results in terms of precision and speed.

2.2   Smile Representation

Using the algorithm proposed above, practitioners imply volatilities from market prices of vanilla options. Two effects are observed:

- **Smile**: For a given maturity, the implied volatilities depend on the strike of the vanilla option.
- **Term-Structure**: For a given strike (moneyness), the implied volatilities depend on the maturity of the option.

These two observations not only demonstrate the inability of the BS model to price and risk-manage financial derivatives while being consistent with the market, but also show the importance of having a good model representing the smile. This is the purpose of the following.

Indeed, the smile representation is one of the key modelling activities of trading every day. A good volatility model is to assume a two-factor Bergomi volatility as described in Bergomi (2015). The two-factor model is the superposition of one fast component with another slow one. We propose here a version that takes less parameters and permits to construct a flexible smile with physical parameters.

$$d\xi_t^T = 2v_{slow}\xi_t^T dW_t^\sigma$$
$$\xi_{0+}^T = \xi_0^T \exp^{v_{fast}X - \frac{1}{2}v_{fast}^2}.$$

where X is $N(0,1)$ and the initial point is random.

We also assume that the spot diffuses with a given Brownian that is correlated with the slow volatility Brownian.

$$dS_t = S_t\sqrt{\xi_t^t}dW_t^S$$

where $\langle dW_t^\sigma, dW_t^S\rangle = \rho dt$.

This model could be named a fast two-factor Bergomi Model. You have indeed the same features as two factors. The interesting fact of doing so is to make it closer to a production level. Besides, it permits to obtain a variety of closed form formula.

1. Vanilla Implied volatility,
2. Volatility Surface deformation,
3. Options on variance swaps,
4. Volatility swaps.

Interestingly, the equivalent local volatility can be obtained using a Monte Carlo based on the volatility process only.

It is given by the following formula as derived in Lee (2001):

$$\sigma_{loc}^2(T, f) = \frac{\mathbb{E}\left(\xi_T^{T2}e^{\frac{-\frac{1}{2}K^2}{(1-\rho^2)\int_0^T \xi_t^t dt}}\right)}{\mathbb{E}\left(e^{\frac{-\frac{1}{2}K^2}{(1-\rho^2)\int_0^T \xi_t^t dt}}\right)}$$

$$\text{where } K(f) = \ln\left(\frac{f}{f_0}\right) + \frac{1}{2}\int_0^T \xi_t^t dt - \rho\int_0^T \sqrt{\xi_t^t}dW_t^\sigma.$$

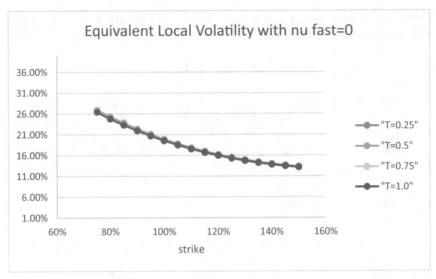

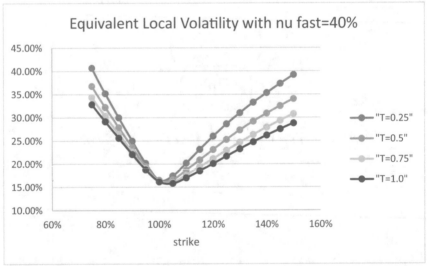

Banks tend to use models they can hedge. Typically they hedge their vega and their delta exposures. Jumps exist in reality and are used as scenarios to estimate the effect on the portfolio of large downward movements. Sometimes, jumps are also used to the marking of the smile surface. Typically, the Variance Gamma model is used for that purpose. The reason is its analytical tractability combined with its parsimony. This is very useful for trading. A good reference on this model is Madan and Seneta (1990).

The evolving environment around smile representation and modelling not only raised various methodological and operational challenges but also theoretical issues. One only has to look at the fast growing literature on the topic to realize how much effort is being put into it. Due to its complexity, smile modelling became a focus for not only practitioners but also for academics. We encourage the reader to consult the following references for further details: Bergomi (2005, 2008, 2009), Carr and Madan (1998).

The next chapter of the book will be dedicated to a model widely used in production nowadays. This is the Local Volatility model. We will introduce only the most important stylized facts and characteristics of the model without giving the mathematical derivations. However, whenever need be, we provide some references for the reader where he can learn about any notions he might not be familiar with.

2.3 Summary of the Chapter: Key Messages

- In this chapter, we **presented a smart methodology** to imply volatility from vanilla prices. The algorithm is based on a **different** implementation of the Black-Scholes formula that is the sum of positive contributions.
- The **proposed** implementation ensures positiveness of the price whichever the discretisation, compared to the classical implementation, which is a difference of two real numbers and shows rounding error problems (especially when the option price is small).
- **This** algorithm has the advantage of being extremely fast, accurate and robust for implying volatility.
- The end of the chapter is devoted to the presentation of a volatility two-factor model in order to construct a flexible smile with physical parameters.

References

Bergomi L (2005) Smile dynamics II. Risk Mag 67–73
Bergomi L (2008) Smile dynamics III. Risk Mag 90–96
Bergomi L (2009) Smile dynamics IV. Risk Mag 94–100
Bergomi L (2015) Stochastic volatility modeling. Wiley
Carr P, Madan D (1998) Determining volatility surfaces and option values from an implied volatility smile. Quant Anal Financ Markets II:163–191
Jäckel P (2006) By implication. Wilmott 60–66
Lee RW (2001) Implied and local volatilities under stochastic volatility. IJTAF 4:45–89
Madan BD, Seneta E (1990) Variance Gamma (V.G.) model for share market returns. J. Bus 63(4):511–524
Press WH, Teukolsky SA (2007) Numerical recipes 3rd edition, the art of scientific computing. Cambridge University Press, New York, NY, USA
Stefanica D (2017) An explicit implied volatility formula. Int J Theoret Appl Finan 20(7)

Chapter 3
Local Volatility Model

Abstract In this chapter, we start first by presenting some typical properties of the local volatility surface, namely: the At-The-Money volatility, skew and curvature. We then move on to the implied volatility dynamics embedded in the local volatility model. We introduce the dynamic of the smile through the Skew Stickiness Ratio (SSR ratio) and lay out a methodology for its computation. The end of the chapter is devoted to the P&L explanation process and the testing of its underlying assumptions (i.e. Delta and Vega hedged positions) through well-chosen scenarios.

Keywords Local volatility · SSR · P&L explain · Hedging hypotheses · Standard theta · Vanilla theta · Exotic theta

The most important objective of a model is to explain the cost of hedging from inception to contract delivery. It is therefore very important at the beginning of the deal to be at least able to match the cost of implementing a portfolio composed of vanilla hedges.

Imagine the trader sells a derivative (say a Put Up and Out) and hedges its position using vanilla options. He might use for that a portion of the final put but also some other contracts including digitals (seen here as a call or put spread) to match the knock out condition.

In this situation, it is essential that the model prices consistently the product (Put Up and Out) and the hedges (portfolio of vanilla options).

In order to fit this minimum requirement in any model, i.e. setting an initial hedge priced consistently, the local volatility model has been proposed. The local volatility model does precisely the cost of the initial hedge. It builds a process that matches all the vanilla option prices and shows a price of the structured product in a consistent way.

This is the best illustration that a derivative model is an elaborate way to extrapolate the present. Indeed, we observe vanilla product prices and by matching these we can price new products which were not initially in the scope.

So let us remember this feature of the local volatility as a big extrapolator of the present.

© The Author(s), under exclusive license to Springer Nature Switzerland AG 2020
O. Kettani and A. Reghai, *Financial Models in Production*,
SpringerBriefs in Finance,
https://doi.org/10.1007/978-3-030-57496-3_3

However, one should acknowledge that this feature is only a minimum and the real value of the model is to transform the premium received at inception into the terminal contract value which is random.

This should not diminish the value of the local volatility model which is used in production all over the city. In the next sections, we shall show how we can augment it in such a way to be able to deliver not only the initial static hedge cost but also an estimation of the future running hedging cost.

After these brief comments on the importance of the local volatility model and its limitations, let us show how we can build it in practice.

Let us assume that the equity process satisfies the following dynamic:

$$\frac{dS_t}{S_t} = \sigma(t, S_t) dW_t \tag{3.1}$$

Assume now that all vanilla options (all strikes and all maturities) are traded and are given by a function, $K, T \to C(K, T)$. This gives the call option prices.

Going from this continuum of market prices to the identification of the model is a classic calibration problem (reverse problem). This particular calibration problem was resolved by Dupire (1994) in his seminal work.

He obtained the following compact and elegant formula:

$$\sigma^2(T, K) = \frac{\partial_T C}{\frac{1}{2} K^2 \partial_{KK} C} \tag{3.2}$$

This formula can be used directly if one applies the same precaution as for the implied volatility calculation (avoiding numerical rounding errors). Alternatively, it is possible to implement this formula using only implied volatilities as shown in Gatheral (2006).

This formula can be expressed solely using directly the implied volatilities Σ_{TK} (cf. Gatheral 2006):

$$\sigma^2(T, K) = \frac{\Sigma_{TK}^2 + 2T \Sigma_{TK} \partial_T \Sigma_{TK}}{1 + 2d_1 K \sqrt{T} \partial_K \Sigma_{TK} + K^2 T \left(d_1 d_2 (\partial_K \Sigma_{TK})^2 + \Sigma_{TK} \partial_{KK} \Sigma_{TK}\right)} \tag{3.3}$$

$d_1 = \frac{\ln\left(\frac{S_0}{K}\right) + \frac{1}{2}\Sigma_{TK}^2 T}{\Sigma_{TK}\sqrt{T}}$ and $d_2 = d_1 - \Sigma_{TK}\sqrt{T}$.

These formulas seem quite involved. It is then time to see what they look like in practice.

We show a typical implied volatility surface, followed by its corresponding local volatility surface.

3.1 Some Stylized Facts About of the Local Volatility

We assume an equity-like underlying with a given volatility surface. Typical properties of this volatility surface are summarized by the following points:

- At the money volatility: This represents the backbone of the volatility surface. It can be thought as a short term initial level, a long term level and a speed to go to the limit (Fig. 3.1).
- At the money skew: This represents the difference in volatility between call volatilities and put volatilities. As for the "At the money" it can represented with the combination of some initial level and some terminal level. However, it is typically a power law decay in time (Fig. 3.2).
- At the money curvature: This represents the difference between the implied volatility slope calculated for puts and calls (Fig. 3.3).

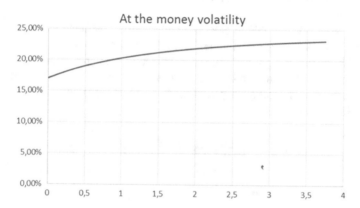

Fig. 3.1 Representation of a typical backbone ATM-volatility

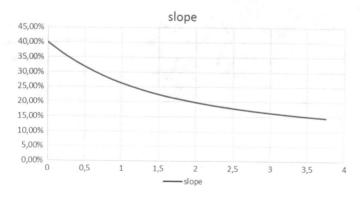

Fig. 3.2 Representation of the at the money volatility slope

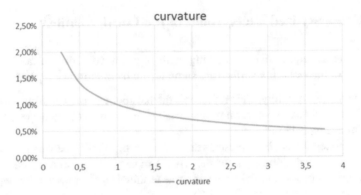

Fig. 3.3 Representation of the at the money curvature

Should you combine these elements of the skeleton you obtain a full volatility surface (Fig. 3.4).

At this stage it is possible to calculate and represent the local volatility surface. The result is shown in the graph under the same scale (Fig. 3.5).

In the following section, we will present some very important properties on the implied volatility dynamics embedded in Local Volatility models. These properties will be put to use in the fourth chapter of the book where we tackle risk-management issues with a LV model.

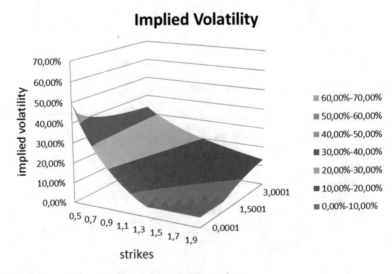

Fig. 3.4 Representation of a full implied volatility surface

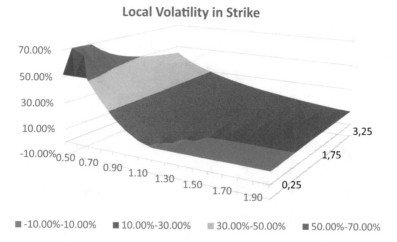

Fig. 3.5 Representation of the local volatility surface

3.2 Other Properties of the Local Volatility

Local volatility models are widely used in the industry. The only source of randomness is the stock price. This makes the calibration of the local volatility easy. In practice, we have an implied volatility surface at pricing time t_0 that matches quoted vanilla options. Using Dupire's equation, we get a local volatility surface, which is used for underlying path generation. This local volatility surface is calibrated at t_0 but there is absolutely no reason for it to remain identical as the spot moves. As a matter of fact, it does change, and quantifying the magnitude of change conveys some interesting information.

3.2.1 The Skew Stickiness Ratio (SSR)

Bergomi (2015) introduced the Skew Stickiness Ratio (SSR) defined for a maturity T as follows:

$$SSR(t_0, T) = \frac{\frac{d\Sigma_{F_T T}^{LV}(t_0, S_{t_0})}{d\ln S_{t_0}}}{\frac{d\Sigma_{KT}^{LV}(t_0, S_{t_0})}{d\ln K}|K = F_T} \tag{3.4}$$

where:

- $\Sigma_{F_T T}^{LV}(t_0, S_{t_0})$ is the At-The-Money-Forward (ATMF) implied volatility, **in the LV model**, with a local volatility surface calibrated at t_0 and a spot value S_{t_0}
- $F_T = \mathbb{E}_{t_0}(S_T)$ is the forward at T seen from t_0

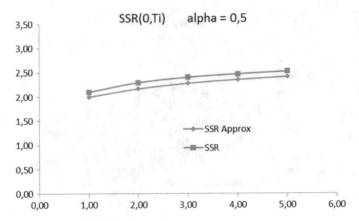

Fig. 3.6 Term-structure of SSR. $\alpha = 0.5$

- $S_{t_0}(T) = \left(\dfrac{d\Sigma_{KT}^{LV}(t_0, S_{t_0})}{d\ln K}\right)_{K=F_T}$ is the ATMF skew at T.

- $\dfrac{d\Sigma_{F_T T}^{LV}(t_0, S_{t_0})}{d\ln S_{t_0}}$ measures the variation of the ATMF implied volatility (in the LV model) when the spot moves.

Bergomi (2015) provides a number of results and approximations for the $SSR(t_0, T)$ in the LV model with a parametric LV surface. We refer the reader to Chap. 2 of his book for a thorough discussion on the results.

In Fig. 3.6, we reproduced the shape of $SSR(t_0, T)$ for different maturities using both an exact computation and the approximation provided in Bergomi (2015). We assumed a local volatility of the form:

$$\sigma_{loc}(t, S_t) = \sqrt{a_0^2 + 2\ln\left(\frac{S_t}{F_t}\right)\left(a_0 a_1 \beta(t) + a_0 a_2 \ln\left(\frac{S_t}{F_t}\right)\right)} \qquad (3.5)$$

where:

$$\beta(t) = \begin{cases} \left(\frac{\tau}{t}\right)^{\alpha} if\ \tau \le t \\ 1\ if\ \tau > t \end{cases}$$

Typical values of the above parameters are:

$$\tau = 0.5(6\,\text{months}), \quad a_0 = 20\%, \quad a_1 = -20\%, \quad a_2 = 3\%, \quad \alpha = 0.5$$

The SSR approximation is given by:

$$SSR(t_0 = 0, T) = 1.0 + \frac{1}{T}\int_0^T \frac{S_{t_0}(t)}{S_{t_0}(T)}\,dt \qquad (3.6)$$

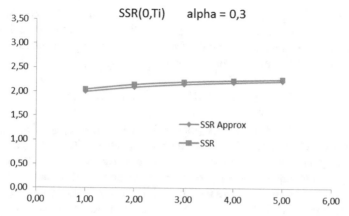

Fig. 3.7 Term-structure of SSR. $\alpha = 0.3$

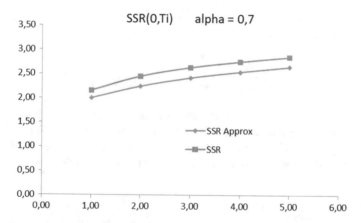

Fig. 3.8 Term-structure of SSR. $\alpha = 0.7$

The term-structure of the SSR is increasing and ranges in the [2, 3] interval. As shown in Fig. 3.7, the approximation is quite good, especially for small values of α (shallow skews) (Fig. 3.8).

We can take the computation of the SSR one step further, and try to see how the ratio behaves over time, i.e. when pricing time is in the future.

3.2.2 The Local Volatility Forward Skew

Let's take a European option with strike K* and maturity T* priced under the local volatility model. As mentioned above, the implied volatility $\Sigma_{K*T*}^{LV}(t_0, S_0)$ is the implied volatility obtained by inverting the option price at pricing time t_0 when the spot value equals S_{t_0}. Assume you priced this option using a Partial Differential Equation method. You can then have the price of this option at any future time t_i and future spot value S_{t_i}. Inverting this price, you get the future implied volatility at strike K*, and maturity T* when the spot value is at S_{t_i}, i.e. $\Sigma_{K*T*}^{LV}(t_i, S_{t_i})$. Mathematically,

$$\Sigma_{K*T*}^{LV}(t_i, S_{t_i}) = BS^{-1}\left(P^{LV}(t_i, S_{t_i}, K_*, T_*)\right) \tag{3.7}$$

where $P^{LV}(t_i, S_{t_i}, K_*, T_*)$ is the LV price at future time t_i ($t_i \geq t_0$) of the European option of strike K_* and maturity T_* (i.e. residual maturity $= T_* - t_i$). S_{t_i} is the value of the underlying at time t_i, and BS^{-1} stands for the inverted Black Scholes function that computes the implied volatility from an option price. The advantage of using a PDE method is that you can have the forward implied volatility on a grid of future times and spots, $(t_i, S_{t_j})_{i=1...n\ j=1...m}$. Thus we have the following values:

$$\Sigma_{K*T*}^{LV}(t_i, S_{t_j})_{i=1...n\ j=1...m}$$

We may define the local volatility forward skew at (t_i, S_{t_i}) as:

$$\frac{d\Sigma_{KT}^{LV}(t_i, S_{t_i})}{d\ln K}\Big|_{K=\mathbb{E}_{t_i}(S(T))=\mathbb{E}(S(T)|S(t_i)=S_{t_i})}$$

Of course, at pricing time t_0, we do not know the value of the underlying at future time t_i. However, we may compute the local volatility skew at (t_i, F_{t_i}) with $F_{t_i} = \mathbb{E}_{t_0}(S_{t_i}) = S_{t_0}e^{(r-q)(t_i-t_0)}$, i.e. the forward at t_i seen at t_0. This is achieved by pricing two options with respective strikes F_T and $F_T * (1 + \epsilon)$, as $E(S(T)|S(t_i) = F_{t_i}) = F_T$. We then get:

$$\frac{d\Sigma_{KT}^{LV}(t_i, F_{t_i})}{d\ln K}\Big|_{K=F_T} = \frac{\Sigma_{F_T(1+\epsilon)T}^{LV}(t_i, F_{t_i}) - \Sigma_{F_T T}^{LV}(t_i, F_{t_i})}{\ln(1 + \epsilon)} \tag{3.8}$$

Figure 3.9 shows the evolution of the forward skew in the local volatility model as time converges towards maturity. Time step is every day from 0 to 4y, meaning that the first point of the plot corresponds to $t_i = 0$ and the final point to $t_i = 4y$ with a time step of 1 day. The initial implied volatility surface is assumed to be parametric as follows:

$$\Sigma_{K_i T_j}^{LV}(t_0, S_0) = \sqrt{a_0^2 + 2\ln\left(\frac{K_i}{F_{T_j}}\right)\left(a_0 a_1 \beta(T_j) + a_0 a_2 \ln\left(\frac{K_i}{F_{T_j}}\right)\right)}$$

where:

$$\beta(T_j) = \left(\frac{\tau}{T_j}\right)^{\alpha} \text{ if } T_j \geq \tau \text{ and } 1 \text{ if } T_j < \tau$$

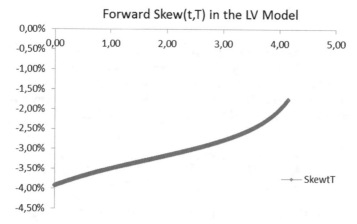

Fig. 3.9 Forward skew in the LV Model. Parameters of the plot: $S0 = 100\%$, $K = 90.48\%$, $r = 1\%$, $q = 3\%$, $T = 5$, Bump $= 1\%$

$$\tau = 0.5(6\,\text{months}), \quad a_0 = 20\%, \quad a_1 = -20\%, \quad a_2 = 3\%, \quad \alpha = 0.5$$

The local volatility map is computed using Dupire formula. In the plot, we retrieve the very well-known fact that local volatility does not preserve the forward skew.

Indeed, the "forward skew" feature is very important when tackling volatility modeling issues. We started first by looking at the BS model, which has the advantage of being very simple and comes with a closed formula for vanilla options. However, the main drawback of this model is its inability to fit the market. So, we needed to improve the modeling in order to fit the market, and we upgraded volatility models from the constant/Term-Structure Black-Scholes volatility to the local volatility. LV models do fit the market, which is a very desirable feature (reproducing the skew and the Term-Structure observed on the market at a certain "present" date). But what happens to the skew as time goes by? In other terms, is the LV model still consistent when pricing forward start options for example, i.e. which depend on the forward skew?

Let us take the example of a call spread (i.e. payoff of the form $1_{\frac{S_{T_2}}{S_{T_1}} \geq B}$) priced at T_0, $T_0 < T_1 < T_2$. The call spread price would depend on the forward skew observed at T_1 for a maturity $T_2 - T_1$. In this section, we showed that one of the main drawbacks of the LV model is that, it doesn't preserve the forward skew. This is why practitioners usually have recourse to stochastic volatility models for a better modeling of the forward skew.

3.2.3   The Local Volatility SSR in the Future

In the previous section, we showed how to compute the forward skew in the local volatility model. This is not enough to compute the SSR ratio. Indeed, we need to measure how the implied volatility changes when the spot does.

Back to our example in the previous section, it is easier to obtain the variation of the implied volatility in the local volatility model at all points $\left(t_i, S_{t_j}\right)_{i=1\ldots n \; j=1\ldots m}$ of the grid.

$$\frac{d\Sigma_{KT}^{LV}\left(t_i, S_{t_j}\right)}{d\ln S_{t_j}} \approx \frac{\Sigma_{KT}^{LV}\left(t_i, S_{t_j}(1+\epsilon)\right) - \Sigma_{KT}^{LV}\left(t_i, S_{t_j}\right)}{\ln(1+\epsilon)}$$

In particular, we may compute the forward covariance in the LV model between the logarithm of the spot and the implied volatility as follows (Fig. 3.10):

$$\frac{d\Sigma_{F_TT}^{LV}\left(t_i, F_{t_i}\right)}{d\ln F_{t_i}} = \frac{d\Sigma_{KT}^{LV}\left(t_i, F_{t_i}\right)}{d\ln K}\Big|_{K=F_T} + \frac{\Sigma_{F_TT}^{LV}\left(t_i, F_{t_i}(1+\epsilon)\right) - \Sigma_{F_TT}^{LV}\left(t_i, F_{t_i}\right)}{\ln(1+\epsilon)} \tag{3.9}$$

We can now compute the SSR ratio at any future time as follows:

$$SSR(t_i, T) = \frac{\dfrac{d\Sigma_{F_TT}^{LV}\left(t_i, F_{t_i}\right)}{d\ln F_{ti}}}{\dfrac{d\Sigma_{KT}^{LV}\left(t_i, F_{t_i}\right)}{d\ln K}\Big|_{K=F_T}} \tag{3.10}$$

We get the following plot (Fig. 3.11).

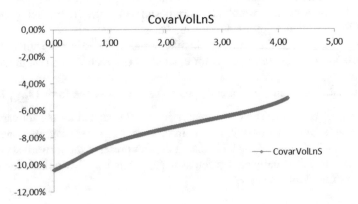

Fig. 3.10 Forward covariance between log-spots and implied volatility in the LV model. Parameters of the plot: S0 = 100%, K = 90.48%, r = 1%, q = 3%, T = 5, Bump = 1%

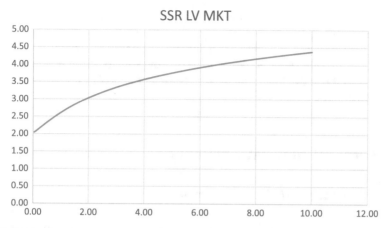

Fig. 3.11 Forward SSR in LV Model. Parameters of the plot: S0 = 100%, K = 90.48%, r = 1%, q = 3%, T = 10, Bump = 1%

The forward SSR in the LV model lies in the [2, 4.5] interval. Although it is not constant, its value remains quite stable over time. It starts to increase when the residual maturity gets small.

3.3 Local Volatility Model: A Pricing Example

Imagine that we would like to price a European product under a given local volatility model. We assume that this product has the following characteristics:

Maturity T,
Payoff ϕ.

For the sake of simplicity we assume no interest rates, repos, nor dividends. Indeed, we can take an affine transformation to take into account repo, rates and dividends as follows: $S_t = a_t X_t - b_t$ where $a_t = S_0 e^{\int_0^t (r_s - q_s)ds} \prod_{t_i < t}(1 - y_i)$ and $b_t = \sum_{t_i} d_i e^{\int_{t_i}^t (r_s - q_s)ds} \prod_{t_i < t_j \leq t}(1 - y_j)$.

Where we have assumed that the stock pays dividends at times t_i equal to $d_i + y_i S_{t_i}^-$ with a repo curve $s \to q_s$ and interest rate curve $s \to r_s$.

Let us then continue with this simple assumption of the absence of dividends, repo and rates and concentrate on the convexity.

The value function satisfies a generalized version of the Black Scholes partial differential equation (PDE) given as follows:

$$\begin{cases} u(T, S) & = \phi(S) \\ \partial_t u + \frac{1}{2}\sigma^2(t, S)S^2 \partial_{SS}u = 0 \end{cases}$$

In the following section, we present one of the most important building blocks of a P&L explanation. Namely, we introduce the P&L evolution equation and explain the first-order hedging hypotheses that simplify the equation. Then, we show how to test these hypotheses in production. Testing the validity of these hypotheses is crucial when it comes to understanding the different impact of market factors on a hedged position.

3.4 Local Volatility PnL Explanation

3.4.1 Derivation of the P&L Equation

Once again, let us remind the reader that the most important objective of a derivative model is its capacity to explain the PnL evolution when used under real market conditions.

In the local volatility framework, without loss of generality we focus our attention to the mono underlying case without dividends, rates nor repos. In this particular case, the pricing function is a function that transforms observable data which persist in the system S_t, Σ_{KT} into a real number $\pi(t, S_t, \Sigma_{KT})$.

The local volatility model is computed on the fly during the pricing process.

The PnL explanation is a four step process which is sequenced as follows:

1. Performing second order Taylor expansion,
2. Making hedging hypothesis,
3. Applying the pricing PDE equation,
4. Applying the recalibration hypothesis.

$$\delta PnL \cong \partial_t \pi \cdot \delta t + \partial_S \pi \cdot \delta S + \partial_\Sigma \pi \cdot \delta \Sigma + \frac{1}{2} \partial_{SS} \pi \cdot \delta S^2$$
$$+ \frac{1}{2} \partial_{\Sigma \Sigma'} \pi \cdot \delta \Sigma \cdot \delta \Sigma' + \partial_{S \Sigma} \pi \cdot \delta S \cdot \delta \Sigma \qquad (3.11)$$

where π is the value of the position seen as a function of respectively time, spot and implied volatility surface respectively noted t, S and Σ_{KT}.

We assume that first orders are hedged $\partial_S \pi = \partial_{\Sigma_{KT}} \pi = 0$.

It is interesting to name these particular hedging situations:

- $\partial_S \pi = 0$ means that the operator is delta hedged,
- $\partial_{\Sigma_{KT}} \pi = 0$ means that the operator is **super vega KT** hedged, i.e. it means that the position becomes insensitive to any movement of the volatility surface. This is an important hypothesis that is difficult to realize in practice. We shall see in the next section what consequence will that hypothesis generate in the PnL evolution equation.

We now use the pricing PDE which stipulates that the pricing function $\hat{\pi}$ is a function of t, S and a given local volatility surface σ:

$$\partial_t \hat{\pi} + \frac{1}{2}\sigma^2(t, S)S^2 \partial_{SS}\hat{\pi} = 0$$

Under the local volatility model, vanilla options are function of time and spot: let us note $\Sigma_{KT}^{LV}(t, S)$ the implied volatility for an option of maturity T and strike K seen at time t and spot S under the local volatility model.

It is important to note that when time moves we keep the implied volatility Σ_{KT} constant and we need to change the local volatility $\sigma_{local}(t, S)$. This is also the case when the spot is bumped to compute the delta and the gamma.

So we can therefore link the derivative of $\hat{\pi}$ with the derivatives of π using the chain rule formula.

$$\partial_t \hat{\pi} = \partial_t \pi + \partial_{\Sigma_{KT}}\pi \cdot \partial_t \Sigma_{KT}^{LV}$$

$$\partial_S \hat{\pi} = \partial_S \pi + \partial_{\Sigma_{KT}}\pi \cdot \partial_S \Sigma_{KT}^{LV}$$

$$\partial_{SS}\hat{\pi} = \partial_{SS}\pi + 2\partial_{S\,\Sigma_{KT}}\pi \cdot \partial_S \Sigma_{KT}^{LV} + \partial_{\Sigma_{KT}\Sigma_{K'T'}}\pi \cdot \partial_S \Sigma_{KT}^{LV} \cdot \partial_S \Sigma_{K'T'}^{LV} + \partial_{\Sigma_{KT}}\pi \cdot \partial_{SS}\Sigma_{KT}^{LV}$$

When we replace the previous quantities in the PnL equation we obtain the following explanation formula:

$$\delta PnL \cong \frac{1}{2}\partial_{SS}\pi \cdot S^2 \cdot \left(\frac{\delta S^2}{S^2} - \sigma^2(t, S) \cdot \delta t\right)$$

$$+ \frac{1}{2}\partial_{\Sigma\Sigma'}\pi \cdot \Sigma\Sigma' \cdot \left(\frac{\delta \Sigma}{\Sigma}\frac{\delta \Sigma'}{\Sigma'} - \sigma^2(t, S) \cdot \partial_{\ln S}\ln\Sigma' \cdot \partial_{\ln S}\ln\Sigma \cdot \delta t\right)$$

$$+ \partial_{S\Sigma}\pi \cdot S\Sigma \cdot \left(\frac{\delta S}{S}\frac{\delta \Sigma}{\Sigma} - \sigma^2(t, S) \cdot \partial_{\ln S}\ln\Sigma \cdot \delta t\right) \tag{3.12}$$

This is the most important result concerning the practical usage of the local volatility model in a real world framework. The derivation seems a bit cumbersome.

3.4.2 Another Simpler Derivation

There is another derivation which relies on the Ito multi-dimensional formula.

Imagine you are looking at a multi-dimensional pricing function $u(t, X_1(t), \ldots, X_n(t))$ with a given set of Ito processes for the underlying $dX_i(t) = \mu_i(t, X_i(t))dt + \eta_i(t, X_i(t))dW_i(t)$ with $\langle dW_i(t), dW_j(t)\rangle = \rho_{i,j}dt$ then the function u satisfies the following multi-dimensional PDE

$$\partial_t u + \sum_{i=1}^{n} \mu_i \partial_i u + \frac{1}{2}\sum_{i,j=1}^{n} \rho_{i,j}\eta_i\eta_j \partial_{ij}u = 0$$

Indeed, the system pricing function $\pi(0, S_0, \Sigma_{KT}(0, S_0))$ satisfies the multi-dimensional Ito which depends on the processes:

$$\frac{dS_t}{S_t} = \sigma(t, S_t)dW_t$$

$$d\Sigma_{KT}^{LV}(t, S_t) = \partial_t \Sigma_{KT}^{LV}(t, S_t)dt + \partial_S \Sigma_{KT}^{LV}(t, S_t)dS_t + \frac{1}{2}\partial_{SS}\Sigma_{KT}^{LV}(t, S_t)\langle dS_t\rangle$$

We can rewrite the last set of equations as follows:

$$\frac{d\Sigma_{KT}^{LV}(t, S_t)}{\Sigma_{KT}^{LV}(t, S_t)} = \mu_{KT}(t, S)dt + v_{KT}(t, S)dW_t$$

where

$$\mu_{KT}(t, S) = \partial_t \ln\Sigma_{KT}^{LV}(t, S_t) + \frac{1}{2}\sigma^2(t, S_t)(\partial_{\ln S \ln S}\ln\Sigma_{KT}^{LV}(t, S_t)$$
$$- \partial_{\ln S}\ln\Sigma_{KT}^{LV}(t, S_t) + \partial_{\ln S}\ln\Sigma_{KT}^{LV}(t, S_t)^2)$$

and

$$v_{KT}(t, S) = \partial_{\ln S}\ln\Sigma_{KT}^{LV}(t, S_t)\sigma(t, S_t).$$

Under the local volatility model we hedge the first order in the general sense (delta but also super Vega KT). Hence, we are not sensitive to the drift $\mu_{KT}(t, S)$. However, we become sensitive to v_{KT}, the volatility of the volatility Σ_{KT}^{LV} and the correlation matrix between the asset and all the volatility points in one hand and the correlation between any option K, T with any other option K', T' on the other hand.

Now applying multi-dimensional PDE satisfied by the value function $\pi(t, S_t, \Sigma_{KT})$, we obtain the following equation:

$$\partial_t \pi + \sigma^2(t, S) S^2 \partial_{SS} \pi + \sum_{KT} v_{KT} \sigma(t, S) \partial_{S\Sigma_{KT}} \pi + \frac{1}{2} \sum_{KT,K'T'} v_{KT} v_{K'T'} \partial_{\Sigma_{KT} \Sigma_{K'T'}} \pi = 0$$

(3.13)

Replacing the term $\partial_t \pi$ in (3.11) gives the PnL equation formula as in (3.12).

What is remarkable in this formulation is the multi-dimensional PDE that the pricing function satisfies. In the next section we shall comment the PnL equation which has the same usual structure as any PnL equation.

3.4.3 PnL Explanation Factors

This equation is quite common, it combines three components:

- **Position** materialized by the triplet $\left(\partial_{SS} \pi, \partial_{\Sigma_{KT} \Sigma_{K'T'}} \pi, \partial_{S\Sigma_{KT}} \pi \right)$,
- **Model** induced parameters $\left(\sigma^2(t, S) \cdot \partial_{\ln S} \ln \Sigma' \cdot \partial_{\ln S} \ln \Sigma \cdot \delta t, \sigma^2(t, S) \cdot \partial_{\ln S} \ln \Sigma \cdot \delta t \right)$,
- **Market** realisation $\left(\frac{\delta S^2}{S^2}, \frac{\delta \Sigma^2}{\Sigma^2}, \frac{\delta S}{S} \frac{\delta \Sigma}{\Sigma} \right)$.

The terms that do not depend on the market realization are deterministic. They contribute to the **theta**. The terms that depend on the realization contribute to what traders call the **carry**.

Fair pricing takes into account the cost of the dynamic hedging.

In a complex market where the volatility surface vibrates depending on the market flows, taking the added cost of such cost is difficult. We shall present in the sequel an elegant and easy way to calculate this effect. Before that, we shall concentrate on the possibility to verify the first order hedging assumptions.

3.4.4 How Do We Test the Hedging Hypothesis in Practice?

The PnL equation relies on two hedging assumptions, perfect delta and super Vega hedge.

In real life, it is easy to test the delta hedging hypothesis. Indeed, one computes the delta of the position and checks that it is zero.

- H1: $\partial_S \pi = 0$,

If H1 is not satisfied, one can easily attribute the PnL variation that is due to the partially delta hedged position.

We can indeed estimate the impact of the PnL variation of a partially hedged portfolio quite simply by measuring the following term:

$$\delta PnL(due\ to\ partial\ delta\ hedge) = \partial_S \pi \cdot \delta S$$

When it come to the Vega KT hedge in real life, testing becomes more difficult. One needs to check the following hypothesis:

- H2: $\forall (K, T)$ $\partial_{\Sigma_{KT}} \pi = 0$

This is a tough operation.

It means that one needs to compute the super Vega of the book and check that it is nil.

This must be performed for every strike and maturity, i.e. compute $\partial_{\Sigma_{KT}} \pi$ and verify that it is zero.

Many reasons make this verification cumbersome and somehow impossible to perform perfectly:

1. It is dependent on the discrete grid of computation, the choice of specific strikes K and maturities T has an influence on the quality of the calculation and its length,
2. It is numerically time consuming even for a reasonable grid. Besides, the bumps that are used for the implied volatility might generate arbitrage during the process of calculation,
3. Even if one can eventually tackle the two previous points at the cost of computation time and the usage of careful and sophisticated techniques, a real book will rely on real tradable and liquid options. Thus, the trader will have to put in place vanilla hedges that are dictated by the market liquidity and as such they differ from the vanilla footprint of the structured product.

All these three points remind us that it is extremely difficult to perfectly hedge the super Vega KT and verify its quality at the book level.

The question in production remains: how can we verify that the position is well super Vega hedged? This question is important in order to estimate the PnL evolution.

So we can have a different perspective. Just like the delta, instead of verifying that the Greek is nil, we can estimate the PnL variation that is due to a non-nil Vega KT.

This change of perspective is essential because it turns a hard if not impossible problem into a feasible task.

The new question becomes how the PnL evolves if the super Vega KT is non-nil?

If we assume that Vega KT is non-nil then the PnL evolves following two different mechanisms.

3.4.4.1 Remarking Mechanism

If $\partial_{\Sigma_{KT}} \pi \neq 0$ it means that the plug of a new volatility surface will impact the value of the position through first order expansion:

$$\delta PnL(due\ to\ remarking\ the\ volatility\ surface) = \partial_{\Sigma_{KT}} \pi \left(\Sigma_{KT}^{new} - \Sigma_{KT}^{old} \right)$$

This value can be checked if a full revaluation of the book is performed with the volatility surface Σ_{KT}^{new} and compare it with the old valuation of the book done with volatility surface Σ_{KT}^{old}.

That is to say:

$$\delta PnL(due\ to\ remarking\ the\ volatility\ surface) \approx \pi \left(\Sigma_{KT}^{new} \right) - \pi \left(\Sigma_{KT}^{old} \right)$$

Here, we just do one extra evaluation of the book with a new volatility surface and compare it with the previous volatility surface.

This implicitly assumes that the second order term is negligible. We can adjust the previous formula using the following expansion by adding the following extra valuation on modified volatility surface in the following way.

$$\frac{\pi \left(\Sigma_{KT}^{new} \right) - \pi \left(2\Sigma_{KT}^{old} - \Sigma_{KT}^{new} \right)}{2}$$

This approach of scenario for market data is extremely powerful as will be seen later in following chapter.

3.4.4.2 Residual Vanilla Theta Mechanism

Just like remarking we would like to compute the theta from the residual vanilla position of the book through well-chosen scenarios. To be more precise, we need to collect both the implicit vanilla Vega coming from client's positions and compare them with explicit vanilla options that are put in place in terms of risk management.

This means that for the structured product one needs to make a partial contribution that highlights the vanilla part of it.

Take a position composed with a call Up & Out option hedged with a terminal call and some digital at the barrier. The hedge is in fact a call option and a call spread (digital part). The theta of the hedge is the natural theta.

The structured product in our situation is a call Up & Out. Its theta mixes both a vanilla part with some part that is exotic. The job is to build a scenario that permits to compute the vanilla part coming from the structured product that we call the implicit vanilla theta.

Computing this residual theta from **explicit (vanilla options used as a hedge)** and **implicit** position (vanilla options embedded within the structured products) can be done quite smartly through a well-chosen scenario [we follow the derivation in

Bergomi (2018)]. We replace the cumbersome exercise of super Vega calculation which can be long and hazardous by a small scenario calculation as described in the following section. Without loss of generality, we assume no drift for the underlying and no discount factor. This makes the formula easy to grasp.

3.5   The Standard Theta

We express the standard theta as a difference between a pricing at time $t + \delta t$ and a pricing at time t keeping the implied volatility $\hat{\sigma}_{KT}$ unchanged:

$$\Theta_{Std} = \frac{d\pi}{dt}\delta t = \pi\left(t + \delta t, S, \widehat{\Sigma}_{KT}\right) - \pi\left(t, S, \widehat{\Sigma}_{KT}\right) \qquad (3.14)$$

To reduce instabilities, we calculate the $t + \delta t$ pricing of $\pi\left(t + \delta t, S, \widehat{\Sigma}_{KT}\right)$ as a t pricing, adjusting the implied volatility to keep the variance unchanged:

$$\pi\left(t + \delta t, S, \hat{\Sigma}_{KT}\right) = \pi\left(t, S, \hat{\Sigma}_{KT}^{*}\right)$$

where $\hat{\Sigma}_{KT}^{*}$ is given by:

$$\hat{\Sigma}_{KT}^{2}(T - \delta t) = \hat{\Sigma}_{KT}^{*}{}^{2}T \Leftrightarrow \hat{\Sigma}_{KT}^{*} = \hat{\Sigma}_{KT}\sqrt{\frac{T - \delta t}{T}} \qquad (3.15)$$

A non-null standard theta is a sign that the hedge is non-static and needs to be rebalanced. It can come from:

- An implicit vanilla theta: there is no real convexity, it can be statically hedged with vanilla options,
- An exotic theta generated by sizeable Gamma/Vanna/Volga.

That is why we split the standard theta into the following sum:

$$\Theta_{Std} = \Theta_{vanilla} + \Theta_{exotic} \qquad (3.16)$$

3.5.1   Decomposition of the Standard Theta: The Exotic Theta

We express the exotic theta as a difference between a pricing at time $t + \delta t$ and a pricing at time t **keeping the vanilla options prices O_{KT} unchanged**:

$$\Theta_{exotic} = \mathcal{P}(t + \delta t, S, O_{KT}) - \mathcal{P}(t, S, O_{KT})$$

where $\mathcal{P}$ represents here the price under a model taking as input the KT vanilla options prices $\pi\left(t, S, \widehat{\Sigma}_{KT}\right) = \mathcal{P}(t, S, O_{KT})$. Note that keeping the vanilla options prices O_{KT} unchanged at time $t + \delta t$ means a special scenario for $\widehat{\Sigma}_{KT}$:

$$\Theta_{exotic} = \pi\left(t + \delta t, S, \widehat{\Sigma}_{KT}^{\star}\right) - \pi(t, S, \widehat{\Sigma}_{KT})$$

With $\hat{\sigma}_{KT}^{\star}$ given by: $\left(\widehat{\Sigma}_{KT}^{\star}\right)^2 (T - \delta t) = \widehat{\Sigma}_{KT}^2 T \Leftrightarrow \widehat{\Sigma}_{KT}^{\star} = \begin{cases} 0 \ if \ T = \delta t \\ \widehat{\Sigma}_{KT}\sqrt{\frac{T}{T-\delta t}} \ else \end{cases}$

As for the standard theta, we can express Θ_{exotic} with t pricing only:

$$\Theta_{exotic} = \pi\left(t, S, \widehat{\Sigma}_{KT}^{\dotplus}\right) - \pi(t, S, \widehat{\Sigma}_{KT})$$

With $\widehat{\Sigma}_{K,\delta t}^{\dotplus}$ given by

$$\widehat{\Sigma}_{KT}^{\dotplus \, 2} (T - \delta t) = \widehat{\Sigma}_{K,\delta t}^{\dotplus \, 2} T \Leftrightarrow \widehat{\Sigma}_{K,\delta t}^{\dotplus} = 0, \widehat{\Sigma}_{KT_i}^{\dotplus} = \widehat{\Sigma}_{KT_i} \ for \ T_i > \delta t \quad (3.17)$$

This exotic theta thus defined measures the actual risk/convexity of the derivative. Indeed if $\Theta_{exotic} \neq 0$, the book is exotic i.e. hedged ratios are not static whereas with a book of vanillas options, Θ_{exotic} is null.

3.5.2 Decomposition of the Standard Theta: The Vanilla Theta

We express the vanilla theta as the difference between the standard theta and the exotic theta:

$$\Theta_{vanilla} = \Theta_{std} - \Theta_{exotic} = \pi\left(t, S, \widehat{\Sigma}_{KT}^{\star}\right) - \pi\left(t, S, \widehat{\Sigma}_{KT}^{\dotplus}\right) \quad (3.18)$$

This vanilla theta thus defined represents the sum of the Black Scholes thetas of the vanillas portfolio which perfectly Vega hedge the derivative:

$$\Theta_{vanilla} = \sum_{K,T} \partial_{\Sigma_{KT}} \pi \frac{1}{Vega_{KT}^{BS}} \Theta_{KT}^{BS} \delta t \quad (3.19)$$

where Θ_{KT}^{BS} is the Black Scholes theta of a vanilla with maturity T and strike K and $Vega_{KT}^{BS}$ is the Black Scholes Vega of the same vanilla option.

So, a book with a non-null $\Theta_{vanilla}$ is not Vega hedged first order.

3.6 Typical Portfolio

A typical portfolio of mono asset in the equity world is constituted of indices (say SX5E, SPX, NKY etc. ...) with a maturity not exceeding 10 years. Besides when the duration is computed at inception of the trades, it is typically 5 years. For the recent years, in the equity world, in a low interest rate environment a best seller product has seen its rise. It is an Autocall product. The client gets a high return if the equity or index rises. To pay for this high coupon the client sells typically a put option to the dealer. This generates the money to compensate for the coupon. This is the point of view of the financial engineering. The trading point of view is hedging this long exposure by selling protection in the market. Therefore, should you look at the market impact of issuing autocalls is to deliver high yield to the client by selling protections on the market.

If we apply the previous theory on a book of mono asset autocalls, we find the following split between the standard theta and the exotic theta.

3.7 Generalize This Approach to Multi Assets and Quanto Effect

In real life, traded options include typically many other effects than the one dimensional option described above. It concerns multi-dimensional possibly with quanto effect and a correlation between the different assets.

The results presented in this chapter, on the Local Volatility model as well as the P&L evolution equation, will serve as a basis for the methodology outlaid in Chap. 4. The objective of the next chapter is to show how complex modeling effects could be taken into account when still pricing a financial derivative with an "inferior model", namely the LV model.

3.8 Summary of the Chapter: Key Messages

- The implied volatility surface is characterized by a number of stylized facts among which we cite:

  - At the money volatility: short-term initial level + long-term level + speed to go to the limit
  - At the money skew: power law decay in time
  - At the money curvature: decay in time

- From the implied volatility surface, and using the Dupire (1994) formula, we get the local volatility surface used for underlying path generation.
- The SSR ratio measures the magnitude of change in the ATMF implied volatility with respect to log-spot variations (in units of the ATMF skew). The term-structure of the SSR is increasing and ranges in the [2, 3] interval. Bergomi (2015) provides an approximation of the SSR that is quite good, especially for shallow skews. In this chapter, we also present a methodology to compute the SSR in the future in the LV model.

- The PnL explanation is a four step process which is sequenced as follows:

  1. Performing second order Taylor expansion,
  2. **Making hedging** hypothesis,
  3. Applying the pricing PDE equation,
  4. Applying the recalibration hypothesis.

  The one equation to remember is:

$$
\delta PnL \cong \frac{1}{2}\partial_{SS}\pi \cdot S^2 \cdot \left(\frac{\delta S^2}{S^2} - \sigma^2(t, S) \cdot \delta t\right)
$$
$$
+ \frac{1}{2}\partial_{\Sigma\Sigma'}\pi \cdot \Sigma\Sigma' \cdot \left(\frac{\delta\Sigma}{\Sigma}\frac{\delta\Sigma'}{\Sigma'} - \sigma^2(t, S) \cdot \partial_{\ln S}\ln\Sigma' \cdot \partial_{\ln S}\ln\Sigma \cdot \delta t\right)
$$
$$
+ \partial_{S\Sigma}\pi \cdot S\Sigma \cdot \left(\frac{\delta S}{S}\frac{\delta\Sigma}{\Sigma} - \sigma^2(t, S) \cdot \partial_{\ln S}\ln\Sigma \cdot \delta t\right)
$$

  P&L is explained by the triplet: second order sensitivities, model induced parameters and market realisations.

- The P&L equation holds if first order sensitivities are hedged perfectly → perfect delta and super-Vega hedges. Testing these hedging hypotheses is very important.
- Testing the super-Vega hedging hypothesis is extremely difficult and time consuming. We provide a smart alternative (i.e. well-chosen scenarios) to test it based on the decomposition of the derivative's standard theta into two components: vanilla theta and exotic theta.

Let's now make sure you have understood this chapter…

Exercise 1

Question 1 Explain why a local volatility model is consistent with the market, compared to a Black-Scholes model.

Question 2 For which type of structured derivatives the local volatility model shows some limits? Why? How practitioners tackle this issue?

Exercise 2

A bank sells an Autocall contract (AC) to an investor. The maturity of the contract is $T_2 = 4y$. The investor and the bank agree on the following:

- If the underlying performance at $T_1 = 2y$ is above 105%, the product is redeemed and the investor gets nothing.
- If the underlying performance at $T_1 = 2y$ is less than 105%, the investor gets the payoff of an ATM put option (i.e. strike = 100%) on the underlying at T_2

Mathematically speaking, the payoff of this autocall writes:

$$\Pi = \left(100\% - \frac{S_{T_2}}{S_{T_0}}\right)^+ 1_{\frac{S_{T_1}}{S_{T_0}} \leq 105\%}$$

The trader in charge of hedging and risk managing this autocall gets the following information from the bank's pricing system:

Data	Value (%)	Pricing	Value
$\Sigma_{T_1}(100\%)$	25	Theta AC	−0.085
$\Sigma_{T_1}(110\%)$	20	Vega AC (T_1, 100%)	0
$\Sigma_{T_2}(100\%)$	30	Vega AC (T_1, 110%)	0
Spot	100	Vega AC (T_1, 105%)	0.072
Interest rate	0	Vega AC (T_2, 100%)	0.70046

The trader has at his disposal (i.e. available on the market) vanilla calls and puts at maturities $T = (T_1, T_2)$ and strikes $K = (100\%, 110\%)$, and decides to hedge the autocall with a European ATM put at T_2 and a quantity Q^* of call spreads (100%-110%) at T_1.

Question 1 Determine the quantity Q^* such that the autocall is globally Vega hedged.

Question 2 Compute the theta of the hedge.

Question 3 Determine the theta Vanilla of this autocall.

Question 4 Estimate the contribution of the Exotic theta to the global theta in this autocall. Is the product fully Vega hedged at first order?

References

Bergomi L (2015) Stochastic volatility modeling. Wiley
Bergomi L (2018) Selected papers. Retrieved from lorenzobergomi.com
Dupire B (1994) Pricing and hedging with smiles. Risk
Gatheral J (2006) The volatility surface, a practitionner's guide. Wiley

Chapter 4
Market Model P&L Explain

Abstract "No LSV model, No problem". How does that sound? Dear reader, leveraging on the P&L equation of Chap. 3, we show how you may still price exotic equity derivatives using the Local Volatility model. This is achieved by adding an adjustment to the price of the derivative in order to account for complex features embedded in the payoff. We focus essentially on the vanna cost, and study the Autocall in its prototypical form, with annual recalls and a European Down-and-In put option at maturity.

Keywords P&L equation · Exotic vanna cost · Autocall payoff

Needless to say, practitioners would prefer a model allowing them to understand their P&L variations rather than predicting the future. Whatever pricing model is used, a trader would have to breakdown its P&L by separating contributions from different effects (markets variables, either observable or not).

Clearly, not all models are equal and do not convey the same information regarding market variables. Although relatively stable and easy to implement, simpler models have the major drawback of lumping together some market effects (for example the covariance between two market variables). On the other hand, putting in production complex, robust, and computationally reasonable models is a daunting task.

One model is particularly popular in the financial derivatives world. This is the local volatility model introduced in the first chapters of this book. However, although local volatility models have the benefit of fitting vanilla options, a number of studies highlighted some of their limits such as not preserving the forward skew and the slope of the volatility term-structure. On the other hand, stochastic volatility models do preserve the forward skew but at the cost of not fitting vanilla options. As a consequence, a new model emerged, namely the LSV model. LSV models fill the gap between both stochastic and local volatility models. They have two degrees of freedom. Although, LSV models require calibration on different instrument classes and calibration may be challenging and time-consuming, the local volatility component ensures that vanilla prices remain unchanged. The "mixing-weight" parameter in the LSV model, usually calibrated to historical data, controls the proportion of

© The Author(s), under exclusive license to Springer Nature Switzerland AG 2020
O. Kettani and A. Reghai, *Financial Models in Production*,
SpringerBriefs in Finance,
https://doi.org/10.1007/978-3-030-57496-3_4

stochasticity in the model. It multiplies the correlation between spot and volatility processes as well as the volatility of volatility. This parameter lies in the interval [0, 1]. When its value goes to zero, we converge towards a local volatility model. When its value goes up towards one, we increase the stochasticity of the model. Introduction to LSV models is not the purpose of this book. The interested reader may find a detailed introduction to the LSV model in some good references on the subject that include but are not limited to (Bergomi 2015).

To make a long story short, quants constantly strive to improve pricing of financial derivatives by enhancing modelling features at a reduced computational cost. In the following, we shall present an innovative approach for taking complex market effects into derivatives prices while still pricing and risk managing with simpler models. Our starting point is the P&L equation (3.11) presented in the section above.

4.1 Theory

4.1.1 Derivation of the Adjustment

Assume we have a derivative π with maturity T that depends on the underlying S_t and the implied volatility Σ. Let's recall the P&L equation (3.11) of an option position between t and $t + \delta t$:

$$\delta PnL \cong \partial_t \pi \, \delta t + \partial_S \pi \, \delta S + \partial_\Sigma \pi \, \delta \Sigma + \frac{1}{2} \partial_{SS} \pi \, \delta S^2$$

$$+ \frac{1}{2} \partial_{\Sigma \Sigma'} \pi \, \delta \Sigma \, \delta \Sigma' + \partial_{S \Sigma} \pi \, \delta S \, \delta \Sigma$$

In practice, traders delta-hedge and Vega-hedge their derivative books to neutralize the risks associated with underlying assets price and volatility movements. Hedging a risk factor requires constant rebalancing of the hedge in order to track the variation in options prices due to the factor's moves. This translates in the P&L equation as follows:

$$\delta PnL \cong \partial_t \pi \, \delta t + \frac{1}{2} \partial_{SS} \pi \, \delta S^2 + \frac{1}{2} \partial_{\Sigma \Sigma'} \pi \, \delta \Sigma \, \delta \Sigma' + \partial_{S \Sigma} \pi \, \delta S \, \delta \Sigma$$

First order derivatives disappear from the equation leaving only second order derivatives, namely: the Vanna, the Volga and the Gamma. Assume we are pricing our derivative using a Local volatility model, the P&L explanation (3.12) equation writes:

$$\delta PnL \cong \frac{1}{2} \partial_{SS} \pi \, S^2 \left(\frac{\delta S^2}{S^2} - \sigma^2(t, S) \, \delta t \right)$$

$$+ \frac{1}{2} \partial_{\Sigma\Sigma'} \pi \ \Sigma\Sigma' \left(\frac{\delta\Sigma}{\Sigma} \frac{\delta\Sigma'}{\Sigma'} - \sigma^2(t, S) \ \partial_{\ln S} \ln\Sigma' \ \partial_{\ln S} \ln\Sigma \ \delta t \right)$$

$$+ \partial_{S\Sigma} \pi \ S\Sigma \left(\frac{\delta S}{S} \frac{\delta\Sigma}{\Sigma} - \sigma^2(t, S) \ \partial_{\ln S} \ln\Sigma \ \delta t \right)$$

In the above, we wrote the P&L variation of our position assuming that the derivative depends on two market variables, namely the spot and the volatility. Of course, one may extend the P&L equation to N market variables by pursuing the expansion of δPnL to the other N-2 market variables. Second order derivatives of the form $\partial_{\alpha_i \alpha_j} \pi$ with $(i, j) \in [1, N]^2$ will appear. These second order derivatives multiply the spread between the realised instantaneous covariance between α_1 and α_2 and their instantaneous covariance embedded in the model (i.e. LV model).

We may now define the carry P&L of the exotic payoff, associated to two market variables α_1 and α_2 as follows:

$$carry_{t_0, T}(\alpha_1, \alpha_2) = \frac{1}{2} \int_{t_0}^{T} \frac{\partial^2 \pi_t}{\partial \alpha_{1_t} \partial \alpha_{2_t}} \left(\left(d\alpha_{1_t} d\alpha_{2_t} \right)_{market} - \left(d\alpha_{1_t} d\alpha_{2_t} \right)_{model} \right) \quad (4.1)$$

Seen from time t_0 (i.e. pricing time), $carry_{t_0, T}(\alpha_1, \alpha_2)$ is a random variable. $\left(d\alpha_{1_t} d\alpha_{2_t} \right)_{market}$ represents the instantaneous market covariance of α_1 and α_2 at future time t, whilst $\left(d\alpha_{1_t} d\alpha_{2_t} \right)_{model}$ is the instantaneous covariance embedded in the pricing model. The carry P&L measures the distortion between model's intrinsic covariances and realized ones over the life of the product.

Most banks do not have the required infrastructure for real-time pricing their whole exotic derivatives book under a complex model (i.e. LSV). Technical equipment (and consequently computing time) is usually the bottle-neck of such task and does not permit the necessary real-time management of the exotic book. In that respect, our idea, which benefitted from fruitful discussions with some market practitioners, including (but not limited to) (Salon 2019), is very simple: to the contractual payoff of the exotic derivative we add a new payoff, i.e. the carry. The carry could be seen as an add-on to the contractual payoff that accounts for complex dynamics. Its price is model dependent if there are no instruments in the market replicating the contractual payoff. Both the contractual and the added payoff are priced and risk-managed under the local volatility model. Ricardo Rebonato famously stated (1999): "*Implied volatility is the wrong number to put in the wrong formula to get the right price*". Likewise, what we are saying here is that: "*we are pricing the wrong payoff, with the wrong model, to get the right price*". The carry acts as a proxy of the exotic effect corresponding to the second derivative $\partial_{\alpha_i \alpha_j} \pi$, embedded in a more complex model. The purpose of our approach is to encode this exotic second order derivative effect in the payoff, in order to be able to price and hedge properly in a real time framework (i.e. with a local volatility model).

In general, most of the exotic impact comes from second order derivatives of the price of the form $\partial_{S\alpha_i} \pi$. α_i could be any other market variable such as implied volatility, repo, dividend... In the following, we will focus mainly on two important market variables: the underlying and the implied volatility. We will study the associated carry P&L and show some numerical applications to vanilla and exotic payoffs (digits, vanilla options and auto callable structures).

4.1.2 Computation of the Second Order Derivative

The second order derivative is given by:

$$\partial_{S_t \alpha_t} \pi_t = \frac{\partial^2 \pi_t}{\partial S_t \partial \alpha_t}$$

α is generally related to a tradable instrument. Let us denote this tradable instrument I_α. For example, implied volatilities are related to option prices, so I_Σ stands for options in this case.

Assume that the trader hedges the impact of the parameter α on his position by buying a quantity q_α of instrument I_α. The hedged position writes:

$$Pos(t) = \pi(t) - q_\alpha(t) I_\alpha(t)$$

$q_\alpha(t)$ is such that:

$$\frac{\partial \pi(t)}{\partial \alpha_t} = q_\alpha(t) \frac{\partial I_\alpha(t)}{\partial \alpha_t}$$

Leading to:

$$\frac{\partial^2 \pi(t)}{\partial S_t \partial \alpha_t} = q_\alpha(t) \frac{\partial^2 I_\alpha(t)}{\partial S_t \partial \alpha_t} + \frac{\partial q_\alpha(t)}{\partial S_t} \frac{\partial I_\alpha(t)}{\partial \alpha_t}$$

We may now derive the second order derivative of the hedged position with respect to S and α as follows:

$$\frac{\partial^2 Pos(t, S_t)}{\partial S_t \partial \alpha_t} = \frac{\partial^2 \pi(t, S_t)}{\partial S_t \partial \alpha_t} - q_\alpha(t, S_t) \frac{\partial^2 I_\alpha(t)}{\partial S_t \partial \alpha_t} = \frac{\partial q_\alpha(t)}{\partial S_t} \frac{\partial I_\alpha(t)}{\partial \alpha_t}$$

The carry between S and α writes:

$$carry_{t_0,T}(S, \alpha) = \int_{t_0}^{T} \frac{\partial q_\alpha(t, S_t)}{\partial S_t} \frac{\partial I_\alpha(t, S_t)}{\partial \alpha_t} ((dS_t d\alpha_t)_{market} - (dS_t d\alpha_t)_{model}) \quad (4.2)$$

4.2 Focusing on the Vanna

In this section, we focus on the term: $carry_{t_0,T}(S, \Sigma)$ with Σ referring to the implied volatility point of interest. We look at the adjustment capturing the exotic vanna cost (exotic vanna times the covariance spread). The adjustment is given by the following equation, where the covariance spread is multiplied by the vanna of the derivative.

$$vanna\,Adjustment = E_{t_0}\left(vanna\,Carry_{t_0,T}\left(S_t, \Sigma_t^T\right)\right)$$

$$vanna\,Adjustment = E_{t_0}\left(\int_{t_0}^{T} \frac{\partial q_\Sigma(t, S_t)}{\partial S_t} \frac{\partial O_\Sigma(t, S_t)}{\partial \Sigma_t^T}\left(\left(dS_t d\Sigma_t^T\right)_{market} - \left(dS_t d\Sigma_t^T\right)_{model}\right)\right) \quad (4.3)$$

With q_Σ defined as in the above and $O_\Sigma(t)$ being the vanilla option price at time t.

To estimate the vanna adjustment, we need to be able to compute three terms: the vanna at all future times up to maturity, the market covariance and the model covariance.

4.2.1 Covariances Spread Computation

In this section, we price our financial derivatives with the local volatility model. For the market part, we assume that a Local Stochastic Volatility model (LSV model) reproduces the market behavior.

Let us denote γ the "mixing-weight" parameter. We define the statistical mixing-weight γ by the following equation:

$$\gamma_{t,T}^2 = 1 - \frac{SSR_{Market}(t, T)}{SSR_{LV}(t, T)} = 1 - \frac{1}{SSR_{LV}(t, T)} \frac{\frac{d\Sigma_{F_t^T T}^{LV}(t,S_t)}{d \ln S_t}}{skew\left(t, F_t^T, T\right)}$$

We may now express the covariance spread in the vanna adjustment thanks to the "mixing-weight" parameter as follows:

$$E_t\left(\left(dS_t d\Sigma_t^T\right)_{market} - \left(dS_t d\Sigma_t^T\right)_{model}\right) = skew\left(t, F_t^T, T\right)(SSR_{Market}(t, T) - SSR_{LV}(t, T))$$

$$S_t \sigma_{loc}(t, S_t)^2 dt$$

with:

$$SSR_{Market}(t, T) - SSR_{LV}(t, T) = -\gamma_{t,T}^2 SSR_{LV}(t, T)$$

$skew\left(t, F_t^T, T\right)$ is the skew of the implied volatility evaluated in $F_t^T = \mathbb{E}(S_T | S_t)$. The vanna adjustment finally writes:

$$E_{t_0}\left(-\int_{t_0}^{T} \frac{\partial q_\Sigma(t, S_t)}{\partial S_t} \frac{\partial O_\Sigma(t, S_t)}{\partial \Sigma_t^T} \gamma_{t,T}^2 SSR_{LV}(t, T) skew\left(t, F_t^T, T\right) S_t \sigma_{loc}(t, S_t)^2 dt\right)$$

(4.4)

If we have only one estimation of the mixing weight parameter, then the relationship above simplifies to:

$$E_{t_0}\left(-\gamma_{t_0}^2 \int_{t_0}^{T} \frac{\partial q_\Sigma(t, S_t)}{\partial S_t} \frac{\partial O_\Sigma(t, S_t)}{\partial \Sigma_t^T} SSR_{LV}(t, T) skew\left(t, F_t^T, T\right) S_t \sigma_{loc}(t, S_t)^2 dt\right)$$

(4.5)

4.2.2 Vanna Computation

To compute the vanna term, one needs to determine first the quantity $q_\Sigma(t, S_t)$. As a reminder $q_\Sigma(t, S_t)$ is defined as the quantity of vanilla options the trader needs to hold to Vega hedge his position. Assume that the position is sensitive to the point Σ_{KT} of the implied volatility surface. $q_{\Sigma_{KT}}(t, S_t)$ is then defined as

$$q_{\Sigma_{KT}}(t, S_t) = \frac{\frac{\partial \pi(t)}{\partial \Sigma_{KT}}}{\frac{\partial O_{KT}(t)}{\partial \Sigma_{KT}}}$$

Using some results in Guennoun (2019), we state that $q_{\Sigma_{KT}}(t, S_t)$, for path dependant contracts subject to some survival condition (for example barrier breaches in autocall features) could be computed as a conditional expectation:

$$q_{\Sigma_{KT}}(t, S) = \mathbb{E}_t(1_{\tau > T} | (t, S_t = S, S_T = K))$$

(4.6)

where $1_{\tau > T}$ is the survival condition, corresponding to arriving at maturity. We outline the important steps of the derivation in the Annex.

In the following, we shall use Eq. (4.6) to compute $q_{\Sigma_{KT}}(t, S_t)$ for the numerical application below. We compute the quantity q for every time t and level of spot S.

4.3 Analysis of a Real Case Example: Case of the Autocall

4.3.1 The Payoff

A standard payoff, widely traded by banks, is the autocall payoff. It has become very popular in the derivatives world and has recently captured a large share of the market. Its popularity is mainly due to the higher returns offered by autocall sellers, in comparison to other structured products without the autocallable feature (i.e. running to their contractual maturity). One may find different variations of the autocall in the market, though in its prototypical form, the payoff is always linked to the performance of an underlying risky asset: index, single stock, basket of stocks…The payoff is made of a set of conditional coupons paid at their respective payment dates, plus a barrier option at maturity. The autocall seller pays the conditional coupons and receives from the buyer a European Down-and-In Put option at maturity. Sometimes, coupons and in-fine option may be based on different types of performances (worst-of, best-of, Asian-in, Asian-out…).

Mathematically, the contract pays at time $T_i (i = 1 \ldots N)$ a conditional coupon defined as follows:

$$\Pi_i = \left(C_i 1_{YB_i \leq P_i} + C_i^{out} 1_{AB_i \leq P_i} \right) \prod_{j=1}^{i-1} 1_{P_j < AB_j}$$

Additionally, if no early redemption occurred the contract pays at maturity N a knocked-in option as follows:

$$\Pi_N = \left(\delta_{type} \cdot (P_N - strike) \right)^+ \left(1 - \prod_{k=0}^{K} 1_{P_k > DB_k} \right) \prod_{j=1}^{N-1} 1_{P_j < AB_j}$$

where:

- AB_i is the auto-callable barrier above which the structure is auto-called at time.
- YB_i is the yeti barrier prevailing at time i.
- P_i is the performance at time i.
- C_i is the coupon paid at time i when the performance is above the yeti barrier and below the auto-callable barrier.
- C_i^{out} is the coupon paid at time i if the structure is auto-called at time i, i.e. the performance P_i above the auto-callable barrier.
- δ_{type} is equal to 1 if the option in-fine is a call option, -1 if it is a put.
- DB_k is the Down-And-In barrier at barrier observation date k. The put option is activated if the underlying performance P_k crosses DB_k (from the above).
- P_N is performance at maturity, the strike being that of the option.

An example of the auto callable payoff cash flows is plotted (Fig. 4.1).

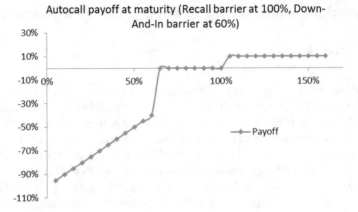

Fig. 4.1 Autocall payoff Cash flows as function of the underlying level

This is an exotic payoff that was responsible for major losses incurred by exotic trading desks over the past years. The payoff shows indeed a strong model dependency due to the cancellation/autocall feature.

In this section, we analyze the vanna adjustment on the following example. Take a 5-year autocall having the following characteristics:

- $AB_i = 100\% \; \forall i \in [1, 5]$
- $C_i = 0 \; \forall i \in [1, 5]$
- $C_{out} = 2\% \; \forall i \in [1, 5]$
- $DB_N = 60\% (N = 5, maturity)$
- $\delta_{type} = -1$ and $K = option\, strike = 100\%$

This autocall pays yearly a conditional coupon of 2%, with a recall barrier at 100%. The option at maturity is a European Down and In Put option, with a strike at 100% and a barrier at 60%. Performances for both the autocall event and the final option are computed the same way:

$$P_i = \frac{S_i}{S_0}, i = 1 \ldots 5$$

where S_0 is the initial value of the underlying and S_i its value at recall date t_i. We choose for simplicity $S_0 = 1$. This choice has absolutely no impact on the following results as we are working with performances for both recall events and the final option, which payoff writes:

$$\Pi_N = \left(K - \frac{S_N}{S_0} \right)^+ = (K - P_N)^+$$

The vanna of the autocall plays an important role in the adjustment. It weights the covariances spread between spots and implied volatility. As we have seen in the above, the autocall payoff could be decomposed into a sum of conditional coupons and a final Down-and-in Up-and-out option (usually a European Down&In Put; PDI). We will study the coupons and the PDI separately, the final vanna adjustment being the sum of all the small adjustments (coupons + PDI).

4.3.2   Vanna Adjustment for the European PDI Up&Out

The autocall seller is long a PDI Up&Out, meaning he receives the following payoff provided no autocall happened before maturity:

$$\left(K - \frac{S_N}{S_0}\right)^+ 1_{\frac{S_N}{S_0} \leq DB_N} \prod_{i=1}^{N-1} 1_{\frac{S_i}{S_0} \leq AB_i}$$

In practice, the trader hedges the Vega of his PDI Up&Out with a European PDI (same maturity, same barrier) without the cancellation effect of the autocall. The price of the PDI is sensitive to one point of the implied volatility surface only: $\Sigma_{DB_N,N}$. To Vega-hedge the sensitivity of his PDI UO to this point of the implied volatility surface, the trader buys a quantity $q_{DB_N,N}$ of the European PDI such that:

$$q_{DB_N,N} = \frac{vega(PDI\ UO)}{vega(PDI)}$$

We have seen in the above how to compute the quantity $q_{DB_N,N}$ as a conditional expectation. The hedged position writes:

$$Pos_N = PDI\ UO - q_{DB_N,N} PDI$$

And the corresponding second order derivative is:

$$\frac{\partial^2 Pos_N}{\partial S \partial \Sigma_{DB_N,N}} = \frac{\partial q_{DB_N,N}}{\partial S} \frac{\partial PDI}{\partial \Sigma_{DB_N,N}}$$

This is the formula used in the vanna adjustment. In the following, we plotted the shape of $q_{DB_N,N}$ as well as the Vega of the PDI as function of the spot (Figs. 4.2, 4.3 and 4.4).

First, few comments on the shape of $q_{DB_N,N}$. When the value of the spot is very low (far from the recall barriers), $q_{DB_N,N}$ is close to one. This is explained by recall events not happening and the value of the PDI UO being close to that of the European PDI. So the quantity of PDI needed to vega-hedge the PDI UO is one. As the spot value increases, $q_{DB_N,N}$ decreases. It converges towards zero for spot values above the recall barriers due to the cancellation feature of the PDI.

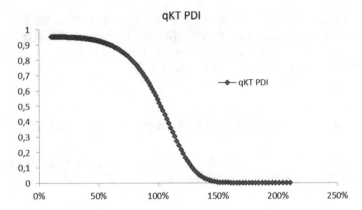

Fig. 4.2 $q(DB_N, N)$ for the PDI. DI barrier at 60%. Maturity $= 5y$. The point of implied volatility of interest is (60%, 5y)

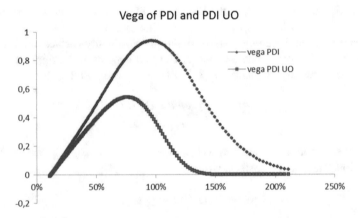

Fig. 4.3 Vega KT of the PDI and the PDI UO with respect to the point of implied volatility (60%, 5y)

The vanna adjustment for the short PDI UO position has a bell shape. It is positive. It reaches its maximum around the recall barriers (for a spot value allowing the Forward to be around the recall barriers, i.e. spot $= 110$ for the numerical example described above). On the wings, the vanna adjustment goes to 0, either because the PDI UO becomes nothing less than a European PDI (left wing), or because the spot value is so high that the cancellation feature has already kicked in before getting to maturity.

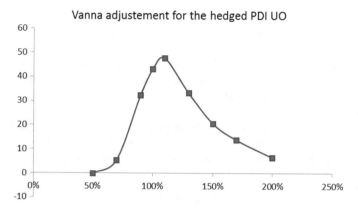

Fig. 4.4 Vanna adjustment for the short 5-year PDI UO position (strike = 100%, barrier = 60%) with annual autocall at 100%. Vanna adjustment is displayed in bps

4.3.3 Vanna Adjustment for the Coupons

We apply the same reasoning displayed above for the PDI UO to the coupons. We assume once again that the trader hedged the Vega of all UO digits with the European digits having the same maturity.

More specifically, the UO digit at the autocall date T_i writes:

$$UO\ Digit_i = C_i 1_{\frac{S_i}{S_0} \geq AB_i} \prod_{j=1}^{i-1} 1_{\frac{S_j}{S_0} \leq AB_j}$$

The corresponding European digit is given by:

$$European\ Digit_i = C_i 1_{\frac{S_i}{S_0} \geq AB_i}$$

The Vega of the European digit is only sensitive to the point (AB_i, T_i) of the implied volatility surface. Thus, we assume that the trader vega-hedges the $UO\ Digit_i$ by buying a quantity $q_{AB_i,i}$ such that:

$$q_{AB_i,i} = \frac{vega(UO\ Digit_i)}{vega(European Digit_i)}$$

where both Vegas are computed by bumping only the point (AB_i, T_i) of the implied volatility surface. Again, in practice, we compute this quantity as a conditional expectation.

The vanna of the hedged position is defined as follows:

$$\frac{\partial^2 Pos_i}{\partial S \partial \Sigma_{AB_i,i}} = \frac{\partial q_{AB_i,i}}{\partial S} \frac{\partial EuropeanDigit_i}{\partial \Sigma_{AB_i,i}}$$

In the following, we plotted the shape of $q_{AB_i,i}$ and the Vega of the $EuropeanDigit_i$ for $T_i = 2y$ as function of the spot. The reader must notice that the first coupon is a vanilla European digit so there is no associated vanna adjustment (Figs. 4.5 and 4.6).

Same conclusions as above apply for the shape of $q_{AB_i,i}$. We also plotted the Vega of the hedge and that of the 2y-coupon Up-and-Out. Vanna adjustments for all the coupons of the 5-year autocall described above are shown in Fig. 4.7.

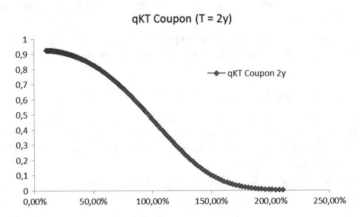

Fig. 4.5 $q(AB_i, i)$ for the coupon UO $T_i = 2y$. Autocall barrier at 100%. The point of implied volatility bumped is (100%, 2y)

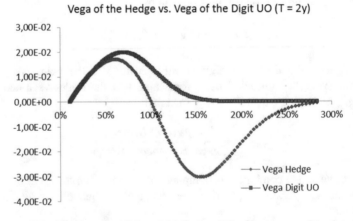

Fig. 4.6 Vega KT of the European digit and the Coupon UO with respect to the point of implied volatility (100%, 2y)

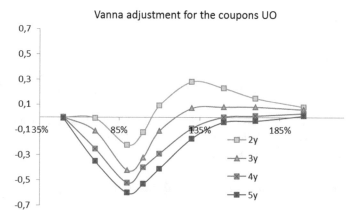

Fig. 4.7 Vanna adjustment for the coupons UO (spot = 100%, autocall barriers = 100%) with annual autocall. Vanna adjustment is displayed in bps

Our first observation is that the adjustment is mainly negative (around recall barriers) for almost all coupons. Second observation is that the adjustment is increasing in absolute value as the number of recall increases (maturity of the digits increasing). Finally, and most importantly, all coupons adjustments are very small compared to the adjustment of the PDI UO (0.5 bps at max compared to 48 bps for the PDI UO).

Upon request, we may provide some pseudo code to help the reader understand how the computation of the vanna adjustment could be implemented in practice. The pseudo code shows the most important steps and is written in a C#/Python format.

4.4   Summary of the Chapter: Key Messages

- In this chapter, we have presented a smart technique to compute adjustments to a local volatility price in order to mimic complex effects embedded in more advanced models.
- We introduced the carry, an add-on to the contractual payoff that accounts for complex dynamics. Its price is model dependent if there are no instruments in the market replicating the contractual payoff. Both the contractual and the added payoff are priced and risk-managed under the local volatility model.
- To compute the vanna carry (focus of the chapter), we need to value the following terms:

  - The second order derivative, i.e. the vanna using a PDE method
  - The covariance spot-volatility spread using the mixing-weight parameter and the Skew Stickiness Ratio (SSR) in the LV model introduced in the Local volatility chapter

- We showed some important results on a "best-seller" financial derivative, namely the autocall payoff for which we found:

  - Adjustments for the coupons are mainly negative (around recall barriers) for almost all coupons, increasing in absolute value as the number of recall increases (maturity of the digits increasing), and very small compared to the adjustment of the PDI UO
  - Adjustment for the PDI UO position has a bell shape. It is positive and reaches its maximum around the recall barriers. Its value goes to 0 on the wings (very small and very high spot values).

Let's now make sure you have understood this chapter…

Exercise
As in Chap. 3, a trader is in charge of risk managing an Autocall contract (AC) on behalf of his bank. The maturity of the contract is $T = 4y$. The investor and the bank agree on the following:

- Annual recall until maturity with the following coupon structure (i.e. the investor gets the following coupons if recall happens)

$$c_{1y}^{out} = 2\% \ c_{2y}^{out} = 2\% \ c_{3y}^{out} = 0\% \ c_{4y}^{out} = 0\%$$

- The recall barriers are $AB_1 = 105\%$, $AB_2 = 120\%$, $AB_3 = 120\%$, $AB_4 = 100\%$
- If recall did not happen before maturity, the bank gets the payoff of an ATM put option (i.e. strike $= 100\%$) on the underlying at T
- The spot value at pricing date is 100%

The bank uses a Local volatility model to price and risk-manage its derivatives book.

Question 1 Explain why the trader should add a fee to the price of the contractual payoff.

Question 2 How many "vanna adjustment terms" does this structure require? What is the sign of the total adjustment? Which part of the contract contributes most?

Question 3 The next day, the spot drops to 80% of its initial level. How is impacted the total vanna cost?

Question 4 :Explain how the total vanna cost is now impacted if the spot rose instead to 140% of its initial level.

References

Bergomi L (2015) Stochastic volatility modeling. Wiley

Guennoun H (2019) Understanding autocalls: real time vega map. SSRN

Salon G (2019) Equity autocalls and vanna negative carries: pricing and hedging with a simple add-on. SSRN

Annex: Derivation of the q_{KT} Expression in the Case of a PDI UO

In this annex, we show the main steps for the derivation of the following equation:

$$q_{KT}(t_0, S_0) = \mathbb{E}\left(1_{\tau > T} | (t_0, S_{t_0} = S_0, S_T = K)\right)$$

Let us start first by recalling the definition of this quantity. $q_{KT}(t_0, S_0)$ is the quantity of vanilla options of strike K and maturity T, i.e. $O(T, K)$, to hold at t_0 when the spot is at S_0, in order to make the portfolio (Exotic-q_{KT} Vanilla) insensitive to the implied volatility point Σ_{KT}.

More generally, in Guennoun (2019), the author wants to find the quantities $\Delta(t, K)$ such that the following holds:

$$\forall (i, j) \in [1, m] \text{x} [1, n], \frac{d}{d \Sigma_{K_i T_j}} \left(\pi_0 - \int_0^T dt \int_0^\infty \Delta(t, K^*) O(t, K^*) dK^* \right) = 0$$

m is the number of strikes of the implied volatility grid and n the number of maturities.

The author states that cancelling the sensitivity to the implied volatility is equivalent to cancelling the sensitivity to the Dupire local volatility points, which writes:

$$\forall (i, j) \in [1, m_{loc}] \text{x} [1, n_{loc}],$$

$$\frac{d}{d\sigma_{loc}(K_i, T_j)} \left(\pi_0 - \int_0^T dt \int_0^\infty \Delta(t, K^*) O(t, K^*) dK^* \right) = 0$$

This gives:

$$\Delta(t, K) = -\partial_t \Gamma(t, K) \, with \, \Gamma(t, K) = \mathbb{E}(\partial_{SS} \mathbb{E}_t(\pi_T | S_t = K))$$

© The Author(s), under exclusive license to Springer Nature Switzerland AG 2020
O. Kettani and A. Reghai, *Financial Models in Production*,
SpringerBriefs in Finance,
https://doi.org/10.1007/978-3-030-57496-3

In our case, our concern is focused on maturity, more precisely the Σ_{KT} point of the implied volatility. Let's say that we want to constitute a portfolio of vanillas of maturities T, $\int_0^\infty \Delta(T, K^*) O(T, K^*) dK^*$, such that

$$\forall i \in [1, m], \frac{d}{d\Sigma_{K_i T}}\left(\pi_0 - \int_0^\infty \Delta(T, K^*) O(T, K^*) dK^*\right) = 0$$

Following the results on the paper, the quantity $\Delta(T, K)$ above is sufficient to satisfy the above equation, and is given by:

$$\Delta(T, K) = -\partial_T \Gamma(T, K) \; with \; \Gamma(T, K) = \mathbb{E}(\partial_{SS}\mathbb{E}_T(\pi_T | S_T = K))$$

When π is a Put Up-and-Out, i.e. $\pi_T = (K - S_T)^+ \prod_{i=1}^{N-1} 1_{S_{T_i} \leq AB_i}$.

We have $\Delta(\mathrm{T}, S) = \Gamma(T, S) - \Gamma(T^+, S) = \partial_{SS}(K - S)^+ \mathbb{E}\left(\prod_{i=1}^{N-1} 1_{S_{T_i} \leq AB_i} | S_T = S\right) - 0$ [Using Eq. 2.5 of the paper Guennoun (2019)]

With $\partial_{SS}(K - S)^+ = \delta(S - K)$, we get:

$$\Delta(T, K^*) = \delta(K^* - K)\mathbb{E}\left(\prod_{i=1}^{N-1} 1_{S_{T_i} \leq AB_i} | S_T = K^*\right)$$

Using this result, we finally get:

$$P_{vanilla} = \int_0^\infty \Delta(T, K^*) O(T, K^*) dK^*$$

$$= \int_0^\infty \delta(K^* - K)\mathbb{E}\left(\prod_{i=1}^{N-1} 1_{S_{T_i} \leq AB_i} | S_T = K^*\right) O(T, K^*) dK^*$$

$$= \mathbb{E}\left(\prod_{i=1}^{N-1} 1_{S_{T_i} \leq AB_i} | S_T = K\right) O(T, K)$$

From which we deduce:

$$q_{KT}(t_0, S_0) = \mathbb{E}\left(\prod_{i=1}^{N-1} 1_{S_{T_i} \leq AB_i} | S_T = K\right)$$

- $O(K, T)$ is the t_0-price of a vanila call of maturity T and strike K
- $\prod_{i=1}^{N-1} 1_{S_{T_i} \leq AB_i}$ is the survival (or non-autocall) of the product from t_0 to maturity T (we can rewrite it: $1_{\tau > T}$) with τ being the survival time.

$q_{KT}(t_0, S_0)$ is the quantity of option O(T, K) to hold at t_0 such as the portfolio (exotic-q_{KT} vanilla) is insensitive to the implied volatility movement.

Bibliography

Berestycki H, Busca J, Florent I (2004) Computing the implied volatility in stochastic volatility models. Commun Pure Appl Math LVII:1352–1373

Berestycki H, Busca J, Florent I (2002) Asymptotics and calibration of local volatility models. Quant Finan 61–69

Bergomi L (2004) Smile dynamics. Risk Mag 117–123

Carmona R, Nadtochiy S (2009) Local volatility dynamic models. Finan Stochast 1–48

Castagna A, Mercurio F (2007) The vanna-volga method for implied volatilities. Risk Mag 106–111

Derman E, Kani I (1994) Riding on a smile. Risk Mag 32–39

Durrleman V (2010) From implied to spot volatilities. Finan Stochast 157–177

Fisher T, Tataru G (2010) Non-parametric stochastic local volatility modeling. In: Global derivatives conference, Paris

Guyon J, Henry-Labordère P (2011) From spot volatilities to implied volatilities. Risk Mag 79–84

Hagan P, Kumar D, Lesniewski A, Woodward D (2002) Managing smile risk. Wilmott Mag 84–108

Lipton A (2002) The vol smile problem. Risk Mag 61–65

Piterbarg V (2005) Time to smile. Risk Mag 71–75

Schweizer M, Wissel J (2008) Term structures of implied volatilities: absence of arbitrage and existence results. Math Finan 77–114

© The Author(s), under exclusive license to Springer Nature Switzerland AG 2020
O. Kettani and A. Reghai, *Financial Models in Production*,
SpringerBriefs in Finance,
https://doi.org/10.1007/978-3-030-57496-3

Index

A
ATMF implied volatility, 20, 35
Autocall, xii, 34, 36, 46, 47, 51, 52

B
Black Scholes, ix

C
Carry, 29, 41, 42, 51
Costs of smile, 5

E
Exotic theta, xi, 33–35

F
Forward skew, xi, 21–24, 39

H
Hedging hypothesis, xi, 26, 29, 35, 36

I
Implied volatility, ix, 8, 11, 43

L
Local volatility, ix, xi, 16, 17, 19, 25, 26, 41, 52, 55

P
P&L equation, xi, 5, 26, 35, 40, 41
P&L explain, xii, 39

R
Remarking mechanism, 31
Risk manage, ix, 2, 7, 10

S
Skew Stickiness Ratio (SSR), 4, 19–21, 24, 25, 35, 51
Smile, xi, 10
Standard theta, xi, 32
Stochastic volatility, ix, 4, 5, 23, 39

V
Vanilla theta, 31–33, 35
Vanna, 43, 44, 46–52
Vanna adjustment, xii, 47, 49, 51

© The Author(s), under exclusive license to Springer Nature Switzerland AG 2020
O. Kettani and A. Reghai, *Financial Models in Production*,
SpringerBriefs in Finance,
https://doi.org/10.1007/978-3-030-57496-3

Printed in the United States
By Bookmasters